SpringerBriefs in Applied Sciences and Technology

SpringerBriefs present concise summaries of cutting-edge research and practical applications across a wide spectrum of fields. Featuring compact volumes of 50 to 125 pages, the series covers a range of content from professional to academic.

Typical publications can be:

- A timely report of state-of-the art methods
- An introduction to or a manual for the application of mathematical or computer techniques
- A bridge between new research results, as published in journal articles
- A snapshot of a hot or emerging topic
- An in-depth case study
- A presentation of core concepts that students must understand in order to make independent contributions

SpringerBriefs are characterized by fast, global electronic dissemination, standard publishing contracts, standardized manuscript preparation and formatting guidelines, and expedited production schedules.

On the one hand, **SpringerBriefs in Applied Sciences and Technology** are devoted to the publication of fundamentals and applications within the different classical engineering disciplines as well as in interdisciplinary fields that recently emerged between these areas. On the other hand, as the boundary separating fundamental research and applied technology is more and more dissolving, this series is particularly open to trans-disciplinary topics between fundamental science and engineering.

Indexed by EI-Compendex, SCOPUS and Springerlink.

Craig E. Banks · Robert D. Crapnell

Screen-Printing Electrochemical Architectures

Second Edition

 Springer

Craig E. Banks
Manchester Metropolitan University
Manchester, UK

Robert D. Crapnell
Manchester Metropolitan University
Manchester, UK

ISSN 2191-530X ISSN 2191-5318 (electronic)
SpringerBriefs in Applied Sciences and Technology
ISBN 978-3-032-10410-6 ISBN 978-3-032-10411-3 (eBook)
https://doi.org/10.1007/978-3-032-10411-3

This Springer imprint is published by the registered company Springer Nature Switzerland AG
The registered company address is: Gewerbestrasse 11, 6330 Cham, Switzerland

If disposing of this product, please recycle the paper.

Preface

Screen-printing is a widely adopted technique for fabricating diverse and functional electrochemical architectures, with applications spanning both academia and industry. It enables the development of next-generation electrochemical sensing platforms and facilitates the translation of laboratory-based innovations into scalable, commercially viable technologies. Despite its widespread use, practical information on how to implement screen-printing effectively in electrochemical applications remains limited. Techniques are often closely guarded, making the process appear opaque or even a "black art" to those new to the field. For practitioners, however, screen-printing offers a powerful and cost-effective route to mass-produce reproducible and electrically robust electrochemical devices.

This *new expanded edition*, volume 2, of the Springer Brief includes updated content, extended chapters, and additional recent literature reports that reflect recent developments in the field. It will be of particular interest to academics, researchers, and industrialists seeking a practical guide, whether starting out in screen-printing, aiming to design and fabricate bespoke electrochemical platforms, or scaling up for commercial production. It also serves as a valuable resource for those with existing experience who wish to deepen their technical expertise and stay current with advancements in this globally adopted technology.

<table>
<tr><td>Manchester, UK</td><td>Craig E. Banks</td></tr>
<tr><td>Manchester, UK</td><td>Robert D. Crapnell</td></tr>
</table>

Competing Interests The authors have no competing interests to declare that are relevant to the content of this manuscript.

Contents

Chapter 1
Introduction and Current Applications of Screen-Printed Electrochemical Architectures

Contents

This chapter offers a general introduction to screen-printed electrochemical platforms, with a focus on notable electrode designs and their reported advantages in the academic literature.

1.1   History of Screen-Printed Electrodes

Within early electrochemical experiments, the utilisation of solid metallic electrodes was a necessity; however, over time, attention has moved towards carbon-based materials within electrochemistry, with the introduction of cheap and affordable approaches such as carbon paste electrodes, which reduce cost significantly. Since the early 1990s, the fabrication of electrode circuits via different printing approaches such as pad-printing, roll-to-roll, and screen-printing have been utilised within electrochemistry. Each of these printing methods offers inherent advantages and disadvantages. For example, pad-printing offers a thin-film transfer

© The Author(s), under exclusive license to Springer Nature
Switzerland AG 2026
C. E. Banks, R. D. Crapnell, *Screen-Printing Electrochemical Architectures*,
SpringerBriefs in Applied Sciences and Technology,
https://doi.org/10.1007/978-3-032-10411-3_1

that can be used in an electrochemical setup; however, this process is not ideal for mass production of electrode systems and therefore has been regularly deemed a pre-requisite to the screen-printing technology. The screen-printing process has the ability for mass production of highly reproducible electrode setups [1–5]. It is with consideration of this that screen-printed electrodes have revolutionised the field due to their capability to bridge the gap between laboratory experiments with in-field implementation.

This is of course compounded by the billion-dollar (per annum) glucose sensing market which has benefited from the incorporation of screen-printed electrodes, since it now allows individuals to be able to measure their blood glucose levels at home, where a result is instantly realised; without recourse to visit a hospital/clinic [6]. This technological approach permits the mass production of highly reproducible electrode configurations that possess excellent scales of economy. Such electrode designs offer improvements in sensitivity, signal-to-noise ratios and reduced sample volumes, producing potential replacements for conventional (solid and re-usable) electrode substrates. Further to this, the ease of the mass production of screen-printed sensors enables their use as one-shot sensors, allowing possible contamination to be avoided, and alleviating the need for electrode pre-treatment as is the case for solid electrodes prior to their use. Figure 1.1 depicts a traditionally used laboratory-based three-electrode system compared to a system which has been screen-printed using conductive inks; such a comparison indicates the ability to create electrochemical setups that are portable, cheap and reproducible.

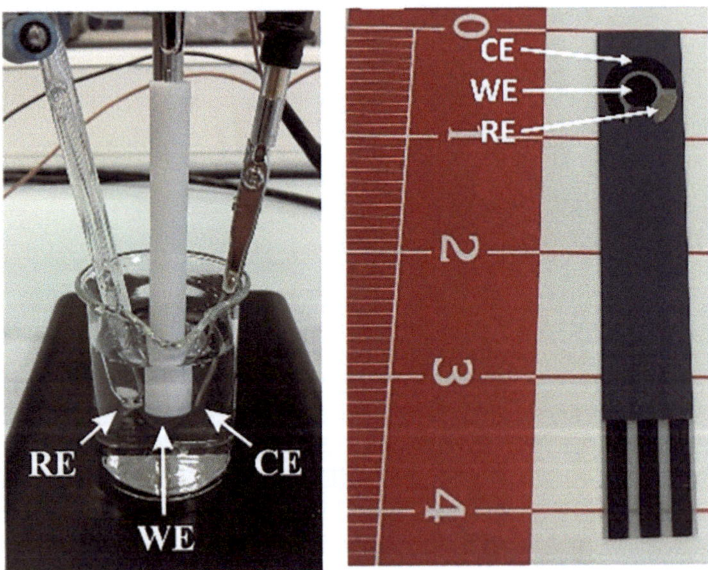

Fig. 1.1 Typical three-electrode system utilising a solid carbon working electrode, saturated calomel reference electrode and a platinum counter electrode (left image). A screen-printed electrode with a carbon composite working and counter electrode and a *pseudo*-Ag/AgCl reference electrode (right image). (Figure reproduced from reference [2]. Creative Commons CC-BY license)

Along with the widely adopted commercially available electrochemical glucose biosensors, screen-printed electrodes have been regularly utilised for a diverse range of other biosensing applications. Their low cost, ease of mass production, and adaptability make them an excellent platform for the integration of different bio-recognition elements beyond enzymes. For instance, DNA-functionalised SPEs have been employed in genosensing for the detection of specific gene sequences, mutations, or pathogens such as *E. coli* and SARS-CoV-2. Aptamer-based SPEs offer a powerful alternative due to their high specificity and stability, enabling the detection of small molecules (e.g., antibiotics, toxins, drugs of abuse) as well as larger biomolecules such as proteins. Similarly, antibody-modified SPEs have been developed for immunosensing applications, allowing sensitive detection of disease biomarkers including cancer antigens and cardiovascular risk markers. These examples highlight the versatility of SPEs as a biosensing platform, extending their role well beyond glucose monitoring to applications in clinical diagnostics, food safety, and environmental analysis.

To further improve the performance of screen-printed electrodes towards applications that require high sensitivity, it can be common to modify the surface further with materials, such as shown in Fig. 1.2. In this method, the material is drop-cast onto the working electrode. It is noted that this method of fabrication is the most common and simplistic technique of modifying a screen-printed electrode surface. However, this approach can potentially offer low reproducibility and often the "films" created are not in fact films. As such, the development of novel inks

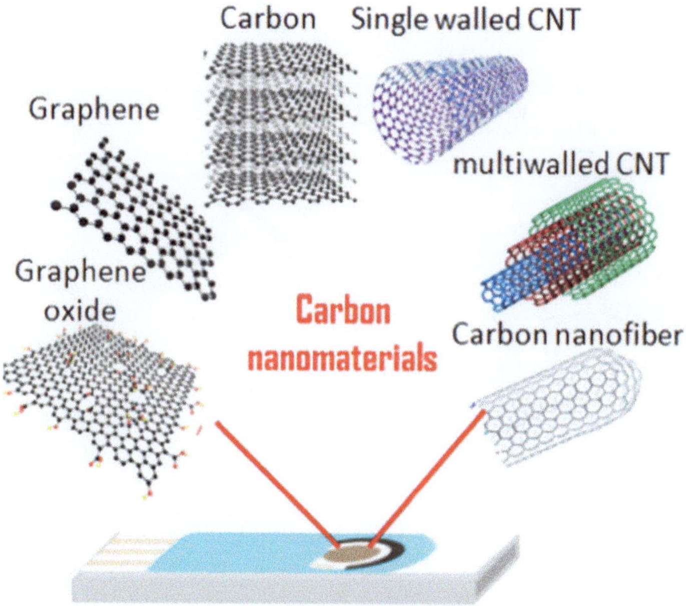

Fig. 1.2 Example of a typical drop-cast fabrication upon screen-printed systems. (Reproduced from Ref. [7])

incorporating the materials is desired, therefore allowing the utilisation of the printing technology and the production of reproducible and consistent electrode surfaces. The ability to create a printing media containing such electrode modifiers gives the user the freedom to produce an even coverage upon the surface.

In addition, screen-printed technology can be regularly applied as potential solar cells with the incorporation of photocatalytic surface groups, in the form of nanoparticles. It is clear within the literature that screen-printing can directly print onto these substrates easily and can create an even surface coverage. Remaining within the area of the renewable energy, focus upon the utilisation of screen-printed electrodes within a potential fuel cell has also been studied by simply screen-printing a minimal amount of the cathode and anode material, potentially reducing the overall fabrication and servicing costs of the future fuel cell technology. Fig. 1.3 shows a screen-printed microbial fuel cell [8].

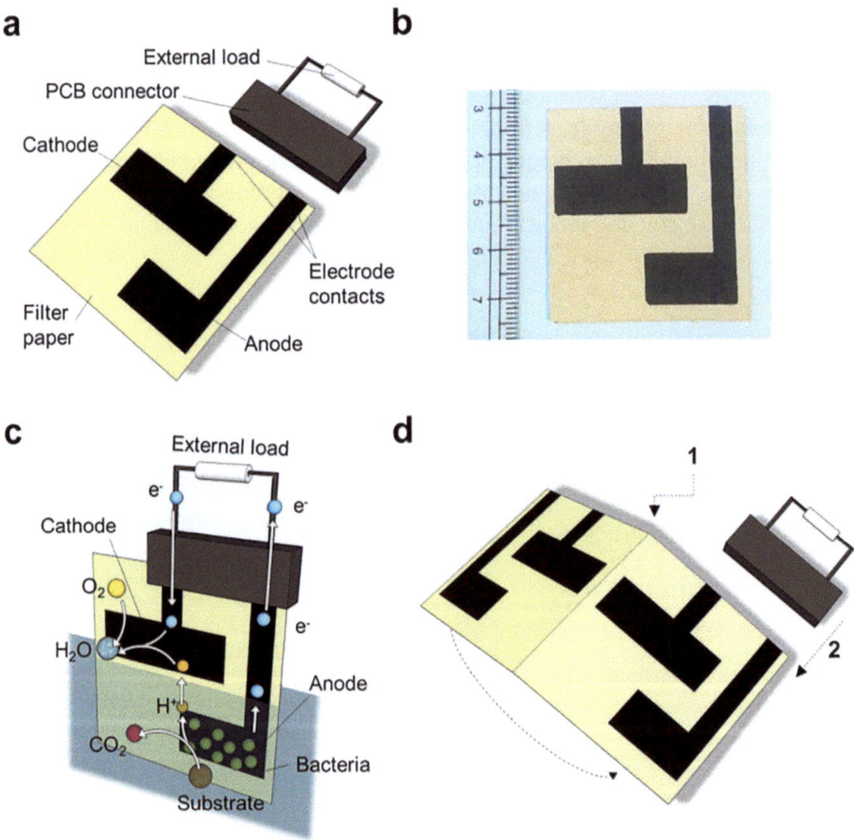

Fig. 1.3 (**a**) Schematic of the paper microbial fuel cell and electrical connection; (**b**) Photograph of the actual paper microbial fuel cell showing size; (**c**) Principle of operation of the paper microbial fuel cell; and (**d**) Assembly of the folded paper microbial fuel cell by folding two paper microbial fuel cell back-to-back (1), with parallel electrical connection (2). (Image reproduced from Ref. [8] Creative Commons CC-BY license)

This work presents the first single-component paper-based microbial fuel cell (pMFC) fabricated by screen-printing carbon-based biodegradable electrodes onto a sheet of paper. Designed as a low-cost, disposable, and environmental-friendly biosensor, the device aims to detect toxic compounds in water, with formaldehyde used as a model contaminant. This approach provides a simple fabrication, where electrodes are printed directly onto paper, with the paper itself serving as a separator and fluidic medium via capillary action, no external pumps, membranes, or potentiostats required. Using two pMFCs folded back-to-back (fpMFC) and electrically connected in parallel resulted in an increased current density fourfold and halved internal resistance compared to a single unit. This approach provides a lightweight, portable, and biodegradable, costing as little as £0.43 per unit with potential for further reduction via scale-up.

Other approaches have reported on the use of screen-printing to produce capacitors. Xu et al. reported the development of screen-printable thin-film supercapacitors using hybrid inks composed of curved nano-graphene platelets (NGP) and polyaniline (PANI), addressing the demand for lightweight, flexible, and integrable energy storage in printed electronics [9]. While NGPs offer high conductivity, surface area, and mechanical robustness, PANI contributes significant pseudo-capacitance but suffers from brittleness and limited conductivity; their combination provides synergistic advantages where NGP acts as a conductive scaffold and PANI enhances charge storage through reversible redox reactions. The formulated NGP/PANI inks, prepared via scalable ball-milling and screen-printed onto conductive carbon fabric substrates, yielded uniform thin films (5–15 μm) with accessible porous structures. Electrochemical evaluation in 1 M H_2SO_4 showed that the optimal 1:1.5 NGP:PANI ratio delivered a specific capacitance of 269 F g^{-1}, a power density of 454 kW kg^{-1}, and an energy density of 9.3 Wh kg^{-1} in two-electrode configurations, alongside excellent cycling stability with no degradation after 1000 cycles. Furthermore, flexible devices maintained good performance (146 F g^{-1}) after 200 bending cycles, demonstrating robustness for wearable and flexible applications. Overall, this work highlights the industrial potential of cost-effective graphene–polymer hybrid inks and screen-printing as scalable routes towards high-performance, mechanically resilient, and fully printable supercapacitors [9].

1.2   Screen-Printed Inks and Designs

Screen-printed inks can be made in the laboratory, or they can be bought commercially. The latter are a guarded secret! Commercially available inks allow one to rapidly translate the screen-printed process into 1000's of electrodes or more. The formulation of conductive inks is not without challenges. The inclusion of polymeric binders, dispersants, or stabilisers, while necessary to ensure printability, adhesion, and mechanical integrity, may simultaneously hinder electron transfer by partially insulating conductive domains. This can reduce electrochemical activity and introduce variability in electrochemical performance. To mitigate these issues and optimise the functional properties of printed electrodes, post-processing treatments are frequently employed. Techniques such as electrochemical activation, laser irradiation, plasma exposure, or mechanical polishing have been shown to remove insulating residues, increase surface roughness, expose active sites, and ultimately enhance electron transfer.

Screen-printed inks are broadly divided into solvent-based and water-based formulations, each with distinct properties and applications. Solvent-based inks use organic solvents such as alcohols, ketones, or esters as the liquid carrier for pigments, binders, and additives. These inks dry rapidly as the solvent evaporates, producing durable, well-adhered films on a wide range of substrates, including plastics, glass, and metals. Because of their strong adhesion, chemical resistance, and robustness, solvent-based systems have traditionally dominated in high-performance applications such as conductive tracks and electrochemical electrodes. However, their reliance on volatile organic compounds (VOCs) poses environmental and health concerns, often requiring extraction systems and stricter handling protocols.

By contrast, water-based inks replace most of the organic solvents with water, sometimes supplemented with small amounts of co-solvent to aid flow and film formation. They are considered more environmentally friendly, emitting fewer VOCs and reducing occupational exposure risks. Water-based inks typically require forced drying, since water evaporates more slowly than organic solvents, and their performance can be influenced by ambient humidity. While adhesion and durability may not always match solvent-based inks, advances in water-dispersible polymers and hybrid formulations have improved their utility, particularly on porous substrates such as textiles and papers.

The choice between solvent- and water-based inks depends on the balance between performance and sustainability. Solvent-based conductive inks containing carbon, graphene, or silver particles offer excellent printability, conductivity, and adhesion to non-porous substrates like PET or glass, making them ideal for sensors and electrodes that must withstand long-term operation. Water-based systems, although less common, are gaining interest in the development of biodegradable, disposable, or eco-friendly devices, where moderate performance may be acceptable in exchange for safer processing and reduced environmental footprint. As such, the shift towards water-based inks reflects broader trends in sustainable manufacturing, while solvent-based inks remain the benchmark for robust electrochemical and functional applications. There are many laboratory-based ink recipes, but these are limited to producing only a handful of electrodes and most, do not scale up, which is rather limiting. That said, we highlight laboratory-based screen-printed inks. As shown in Fig. 1.4a, a new conductive ink has been prepared by adding carbon black with poly(vinyl alcohol) (I). The adhesive mask was glued to a PET substrate, and the conductive ink was spread (II) on the surface. The excess ink and adhesive mask were removed (III), and the electrodes were properly separated (IV).

After applying silver ink on the reference electrode and clear nail polish to delimit the electrode system area (V), the sensor was ready to use. Furthermore, Fig. 1.4b, shows the preparation of the αsyn-BSA-Aαsyn-Ga-Cys-Pd/CB-PVA biosensor. On the surface of CB-PVA, Pd nanoparticles were added (I). Afterward, the device underwent sequential modifications of cysteine (II), glutaraldehyde (III), and anti-α-synuclein (IV). The α-synuclein was then added to the system (V), and the analysis was performed (VI). Wang et al. [11] indicates that the choice of carbon ink

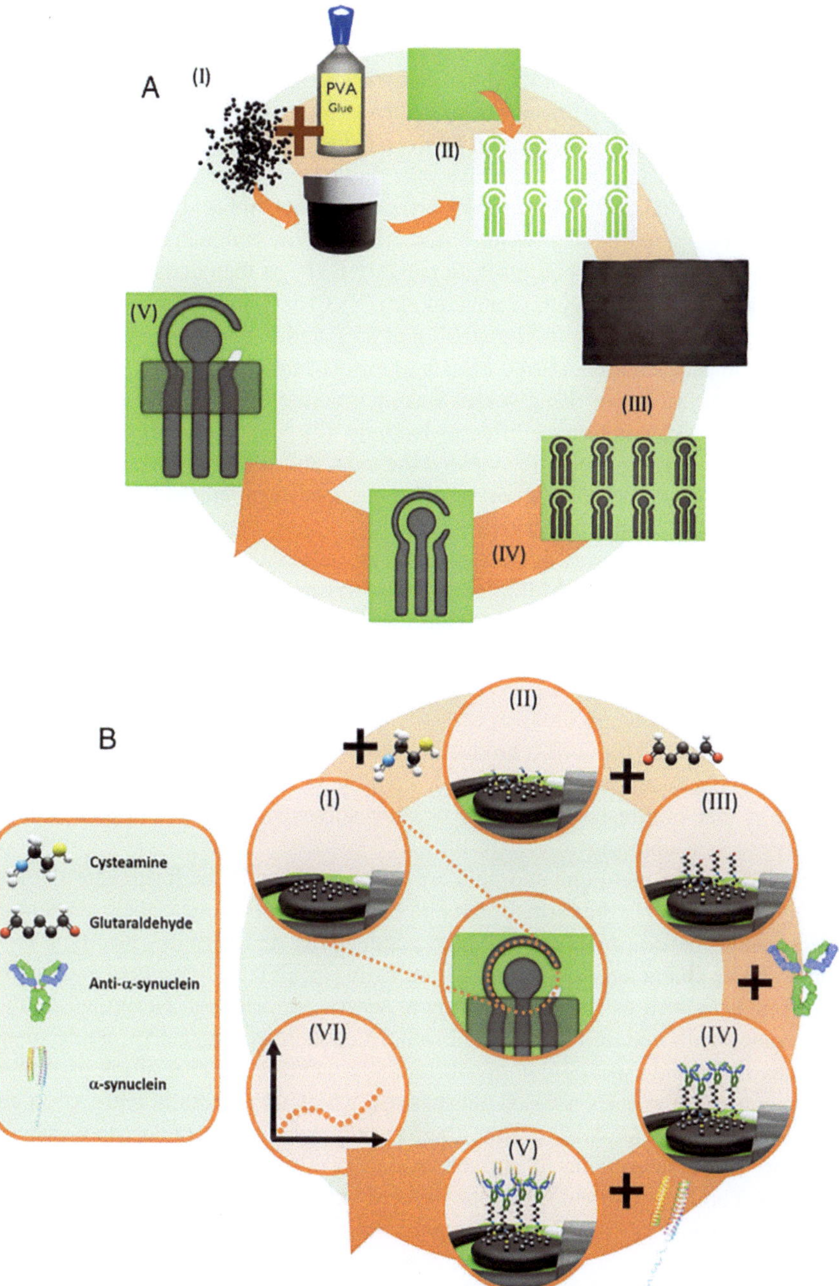

Fig. 1.4 Production of the screen-printed electrode using carbon black-poly(vinyl alcohol) and a step-by-step approaches (**a**) and the approach to modify with palladium nanoparticles (I), (II) cysteine, (III) glutaraldehyde, (IV) anti-α-synuclein, and (V) α-synuclein and electrochemical analysis by electrochemical impedance spectroscopy (VI) (**b**). (Figure produced from reference [10])

should rely upon the application at hand, due to the differences within the electrochemical signals seen with a range of inks.

The use of water-based inks helps improve the sustainability of screen-printed platforms since they minimise or remove toxic solvents, and use low-cost, biodegradable polymers such as chitosan. Derived from chitin, chitosan offers significant advantages for sustainable conductive inks due to its biocompatibility, biodegradability, and excellent film-forming properties. As shown in Fig. 1.5, one can see how Camargo et al. prepared their water-based ink [12]. The conductive water-based ink was formulated using carbon black Super P (CBSP) as the conductive filler and a biopolymer matrix of poly(vinylpyrrolidone) (PVP) and chitosan. First, the polymer solution was prepared by dissolving 1.0 g of PVP and 1.0 g of chitosan in 50 mL of a 2% acetic acid solution, stirred for 2 h to ensure homogeneity. To improve adhesion and film flexibility, 3% glycerol was incorporated. CBSP was then dispersed into this polymer solution at different loadings (3, 5, and 7 wt%) and stirred for 24 h. The 5% CBSP formulation yielded the most uniform, reproducible, and stable ink, avoiding agglomeration observed at higher concentrations and the poor screen-printing quality found at lower ones.

The optimised ink was deposited using an automated screen-printing process onto polyester substrates. Multiple squeegee passes were investigated, with three passes providing the most homogeneous and conductive layers. After printing, the electrodes were dried at 60 °C for 30 min between layers, followed by a final curing at 60 °C for 12 h. A silver/silver chloride paste was applied to define the reference electrode, and a dielectric ink was used to insulate contacts and delimit the working area. This method produced reproducible electrodes with good adhesion, stability, and electrochemical performance, while employing a sustainable, water-based formulation that eliminates hazardous organic solvents and utilises biodegradable binders such as chitosan.

Silver, copper, platinum, and gold are widely utilised in the formulation of conductive inks, often in the form of metallic nanoparticles, owing to their outstanding electrical properties, most notably high conductivity and very low electrical resistance. These characteristics make them attractive candidates for printed and flexible electronics as well as for electrochemical sensor applications. Nevertheless, their use comes with certain limitations. From an economic perspective, noble metals such as silver and especially gold substantially increase the cost of ink production when compared to carbon-based alternatives, which are generally more affordable and scalable. Furthermore, despite their excellent intrinsic conductivity, metallic inks are more susceptible to surface oxidation during storage or operation. For instance, copper is particularly prone to rapid oxidation, while even silver and gold can undergo surface modifications under certain environmental or electrochemical conditions. Such processes can hinder charge transfer kinetics at the electrode–electrolyte interface, thereby reducing the electron transfer rate and negatively impacting the sensitivity, stability, and overall analytical performance of the sensor. Such other examples of printable inks, include the fabrication of platinum and gold screen-printed sensors which have been applied towards the electroanalytical sensing of chromium species (VI and II) in the case of the gold sensor, and both hydrazine and hydrogen peroxide for the case of the platinum sensor [13, 14]. Crucially,

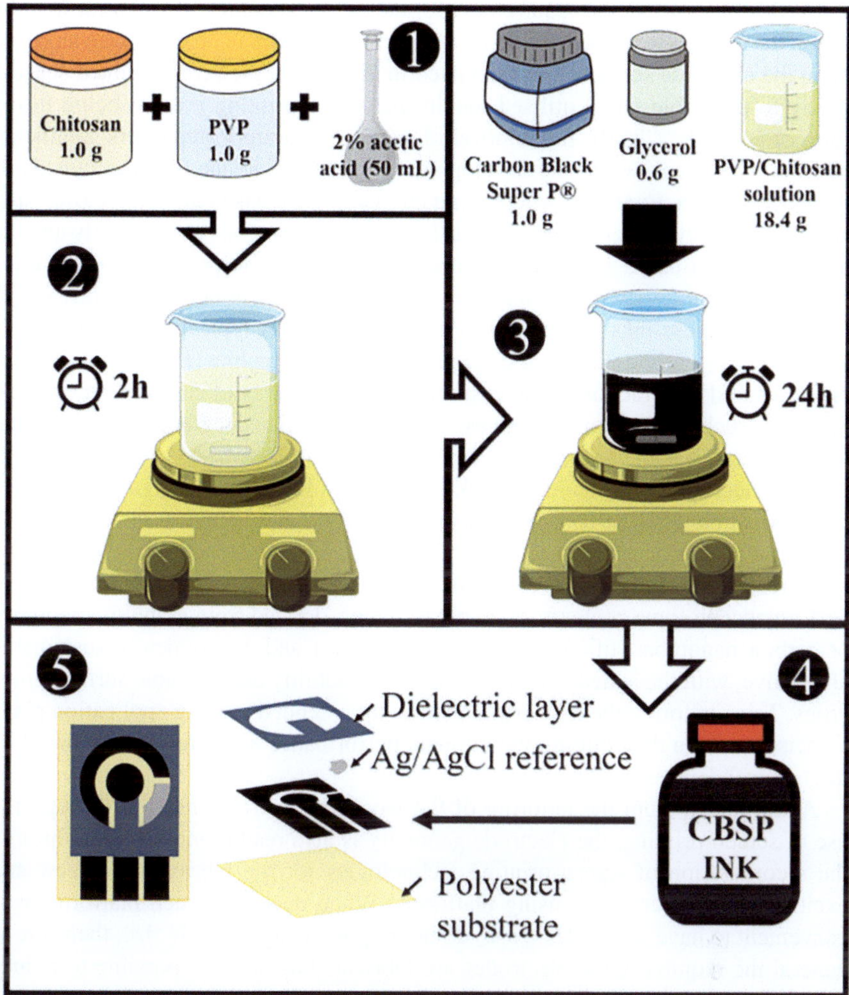

Fig. 1.5 Scheme of SPE preparation using the carbon black modified with Super P. (1) To create the polymer solution, 1.0 g of PVP, 1.0 g of Chitosan, and 50 mL of a 2% acetic acid solution were used, (2) The mixture was mixed for 2 h with the assistance of a stirrer. (3) To create the ink CBSP, 1.0 g of CBSP, 0.6 g of glycerol and 18.4 g of the polymer solution were mixed for 24 h with the assistance of a stirrer. (4) The ink was ready to be used. (5) The screen-printing machine was used to perform the SPE system, the squeegee was passed over the screen three times to ensure an even surface for the carbon ink. This carbon ink was applied to a polyester substrate and dried at 60 °C for 30 min. Ag/AgCl paste was then applied to form the reference electrode layer and dried at 60 °C for 30 min. To define the three electrodes and delineate the sample area, a dielectric ink was used to cover the terminals. The finished sensor was dried at 60 °C for 12 h to ensure it was ready for use. (Figure produced from reference [12]. Creative Commons CC-BY license)

it was determined that the requirement for electrode potential cycling prior to utilisation (as is the case for bulk noble metal macro electrodes in order to form an oxide upon the electrode surface) was alleviated in the case of the screen-printed sensors owing to the noble metal utilised within the screen-printing process being in the form of an oxide. Clearly, the removal of such a preparatory step offers significant benefits when considering the development of sensors intended for use outside of the laboratory environment, where rapid and facile analysis is imperative (one-step analysis). However, mentioned previously the metallic options can be costly and the need for carbon-based materials has become necessary. In response to this, the utilisation of graphitic based inks for an assortment of applications has been realised. Not only do these inks replicate the electrochemistry of typically used electrodes such as glassy carbon (GC) and edge plane pyrolytic graphite (EPPG) but they are manufactured at a fraction of the cost. In addition, an array of carbon materials can be used, such as carbon nanotubes, graphene, mediated carbon structures and nanoparticles to name a few.

The selection of the conductive ink composition is a critical design parameter in the fabrication of electrochemical sensors. Beyond conductivity, factors such as resistance to oxidation, long-term stability, printability, mechanical flexibility, and cost must be carefully balanced. While metallic inks may offer superior electrical performance in some contexts, carbon-based inks, derived from graphite, graphene, or carbon nanotubes, often provide a more robust and economically sustainable alternative, with the added benefits of chemical stability and tuneable surface properties. Thus, tailoring the ink formulation to the intended sensing application plays a decisive role in determining the ultimate performance and reliability of the electrochemical device.

Additionally, from the tailoring of the working electrode material through the use of screen-printing, the electrode geometry is also readily manipulated through the incorporation of screen-printing technologies for the fabrication of new and exciting electrochemical sensing platforms. When designing such platforms it is convenient to have electrodes which do not require a large sample size, therefore in general the required three electrodes are fabricated as close as possible to reduce resistance which can potentially distort the measured voltammetry; the most common electrode configuration is the 3 mm working electrode with on-board reference and counter (see Fig. 1.1). In addition to macroelectrode architectures, microelectrodes can be designed and fabricated to mimic that of true microelectrode behaviour (i.e., near elimination of contributions from planar diffusion rather than solely radial/convergent diffusion) and further enhancements in the electroanalytical sensing performances. The conversion from macro- to microelectrode systems within electrochemical applications possess some significant advantages, such as: smaller diffusion layer, greater mass transport, lower transient current and more resistive to the effects the ohmic drop. These systems also have benefits within the fabrication method, due to the minimal ink utilised during screen print.

A variant on the traditional co-planar disc-shaped microelectrode which continues to gather further interest within current literature is the microband electrode. Band electrodes are fabricated to be macroscopic in-length but microscopic in width. These electrode configurations have been reported to offer the additional advantage of allowing larger currents to be obtained compared to a microdisc due to the increased electrode area, while the width of the band is still maintained in the micrometre range to ensure convergent diffusion is still dominant. Generally, these have been fabricated via the screen-printing method, however the utilisation of gold and platinum inks are generally printed upon a ceramic substrate, offering less user-ability and applicability within-the-field. Upon this printed electrode, a Pyrex glass slide is used to define a thin microband system [15, 16].

An adapted approach within electroanalytical applications has been carried out, see Fig. 1.6, where one can observe how ultra-microband electrodes are produced

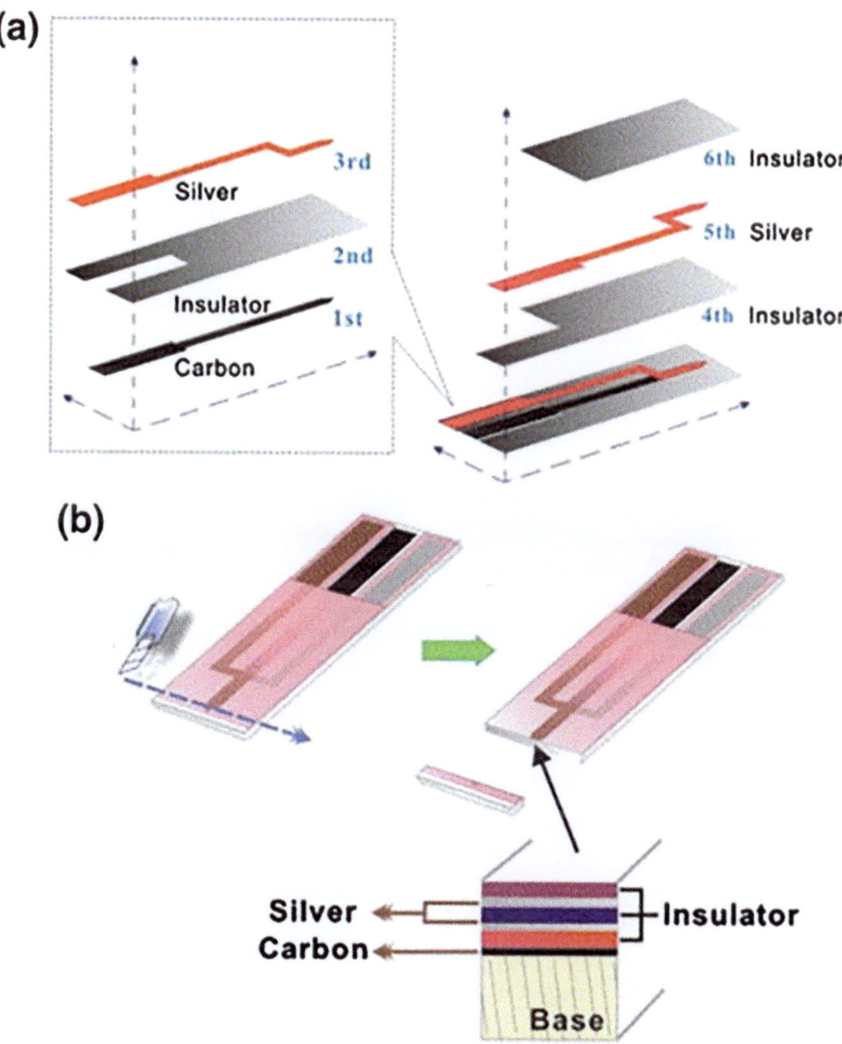

Fig. 1.6 (**a**) A schematic of a layer-by-layer structure of the screen-printed ultra-microelectrode assembly. (**b**) Cross-sectional diagram the electrode system with a built-in three-electrode configuration. (Reproduced with permission from Ref. [17])

using a cutting machine. To alleviate the issue that a range of sizes will be realised using the cutting machine, Metters et al. [18] reported the use of a V-Mesh (Vecry Mesh) (explained further within Sect. 2.1) for a defined and reproducible screen-printed 50 μm (width) microband electrode, such electrode was employed as an electroanalytical sensor.

Other useful approaches have been realised using screen-printing technology, for example, the production of a screen-printed array (SPA), containing six carbon working electrodes, for the detection of procalcitonin (PCT), an important biomarker for sepsis in undiluted human serum samples; see Fig. 1.7 [19]. Screen-printed arrays are highly useful because they combine the low cost, reproducibility, and scalability of screen-printing with the advantages of parallel and multiplexed electrochemical measurements. By fabricating multiple electrodes with consistent geometry and surface area on a single platform, they enable reproducible results and high throughput testing while requiring only small sample volumes. Different working electrodes within the array can be individually modified, allowing simultaneous detection of multiple analytes or comparative evaluation of electrode materials under identical conditions. Since these arrays often integrate reference and counter electrodes, they provide a self-contained, portable, and disposable sensing platform. Their versatility, ease of customisation, and compatibility with automation make them valuable for applications ranging from point-of-care diagnostics and environmental monitoring to electrode screening for electrochemical reactions

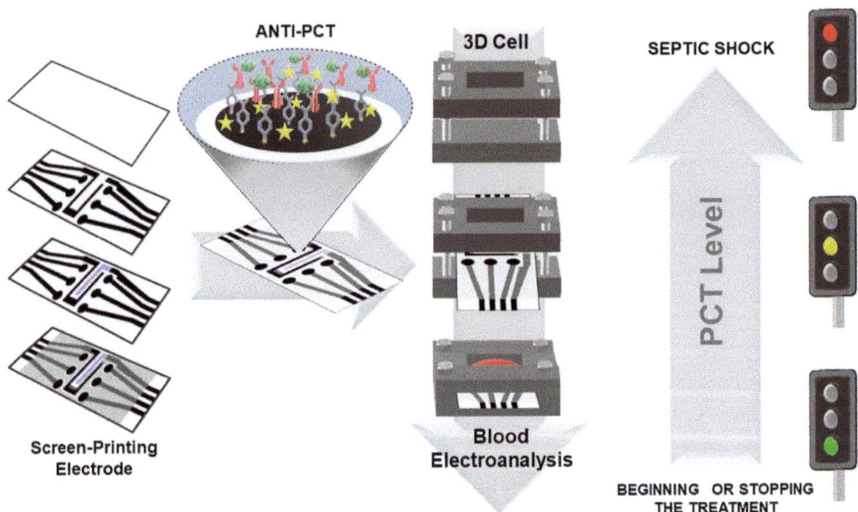

Fig. 1.7 Schematic representation of the screen-printed array production and connection to 3D-printed sample holder. (Figure reproduced from [19]. Creative Commons CC-BY license)

1.2.1   Screen-Printed Designs [7]

In the use of the SPAs shown in Fig. 1.7, three electrodes were modified for the PCT sensing, with three others kept as internal controls. When running controls and tests on the same chip, one can directly compare responses under identical conditions. This eliminates variability due to electrode-to-electrode differences, handling errors, or matrix effects from serum providing an internal validation.

The screen-printing methodology produced SPAs with six individual working electrodes that exhibit an inter-array reproducibility of 3.64% and 5.51% for the electrochemically active surface area and heterogenous electrochemical rate constant respectively [19]. The SPAs were modified with antibodies specific for the detection of PCT, where each stage simply uses droplets incubated on the surface, allowing for their mass production. As shown in Fig. 1.8 the modification process began with diazonium chemistry applied through a droplet method. A droplet containing 4-aminobenzoic acid and sodium nitrite was deposited on the graphite working electrode surface, followed by ascorbic acid. This resulted in the covalent attachment of a carboxyl-terminated aryl monolayer. Unlike conventional electrodeposition, which often produces a dense and blocking layer that reduces electrochemical activity, this droplet-based method provided a more controlled modification that maintained good electrode performance.

Following the formation of the carboxyl-functionalized surface, EDC/NHS coupling chemistry was employed to activate the carboxyl groups, creating reactive ester sites suitable for antibody attachment. A droplet containing procalcitonin-specific antibodies was then applied, allowing covalent immobilisation of the

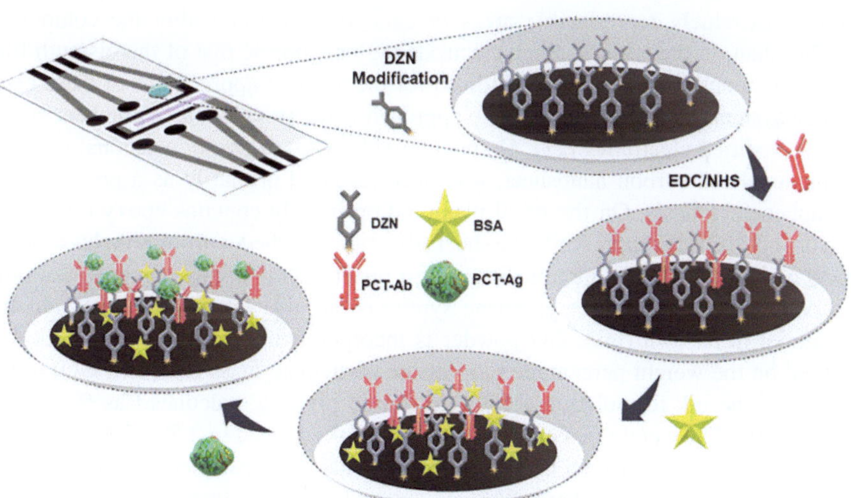

Fig. 1.8 Schematic of the biosensor droplet modification process. (Figure reproduced from [19]. Creative Commons CC-BY license)

biomolecules onto the electrode surface. This ensured strong and stable attachment of the recognition element required for selective PCT detection. To minimise non-specific interactions, the unreacted sites on the electrode were passivated using a droplet of bovine serum albumin (BSA). This blocking step prevented background signals and enhanced the selectivity of the biosensor. Each modification stage was validated using electrochemical impedance spectroscopy, where the observed increase in charge transfer resistance confirmed the successful stepwise functionalization of the electrode surface. The resulting biosensor demonstrated reliable performance, with a linear dynamic range between 1 and 10 ng/mL, a limit of detection as low as 0.7 ng/mL, and good reproducibility (intra-day variation of 4.0% and inter-day variation of 5.05%). Tests in human serum showed selective detection of PCT without interference from common inflammatory biomarkers such as C-reactive protein (CRP) and interleukin-6 (IL-6). Overall, this droplet modification strategy offered a simple, scalable, and internally validated approach to producing biosensors that could be easily adapted for point-of-care testing in sepsis diagnosis.

1.3   Bulk Modified Screen-Printed Electrodes

Consider the use of bismuth oxide which is a common oxide form that can be mixed within a printable ink for detection of metal ions, with such modification being *akin* to that of highly toxic mercury electrode systems. During analysis, bismuth oxide can be electrochemically reduced to metallic bismuth, which enhances sensitivity in stripping voltammetry by forming alloys with trace metals such as lead, cadmium, and zinc. Unlike that of an in situ or ex situ modification the fabrication process can create extremely reproducible areas of catalytic material within the composite, rather than on the surface. Such electrode systems mimic that of the bismuth film electrodes and therefore can be utilised for the array of applications of these electrode systems [20]. Another notable approach is the production of graphene oxide bulk modified electrodes [21]. Graphene oxide (GO), a two-dimensional oxygen-functionalized carbon nanosheet, was once regarded primarily as a precursor for graphene synthesis. On the basal plane, GO primarily contains epoxy (–O–) and hydroxyl (–OH) groups, while the sheet edges are enriched with carboxyl (–COOH) and carbonyl (–C=O) functionalities. As a result, GO is highly versatile, not only serving as a precursor for graphene synthesis but also functioning as a valuable material in its own right. GO powder is incorporated into the bulk graphitic ink based on the weight percentage ratio of the particulate mass (MP, i.e., GO) to the mass of the ink formulation (MI) used during printing, calculated as $\% = (MP/MI) \times 100$. The weight percentage of MP to MI is adjusted to 2.5%, 5%, 7.5%, and 10%, yielding four distinct GO-modified inks that are subsequently screen-printed onto the working area of bare SPEs. As shown in Fig. 1.9, one can see the electroanalytical response towards dopamine and the various electrodes. Note that the 10% GO-SPE shows a larger response than the bare SPE, where the oxygenated species present on GO facilitates the oxygenated electrocatalytic reactions.

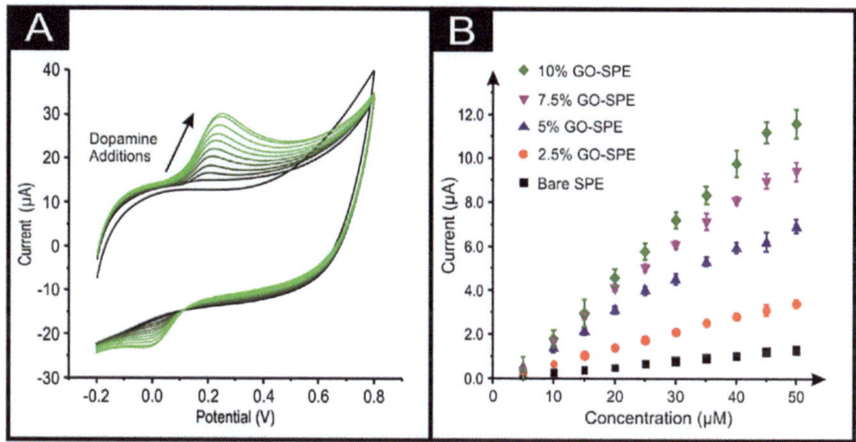

Fig. 1.9 (**A**) Typical cyclic voltammetric response obtained utilising 10% GO-SPEs by sequentially adding aliquots of DA into pH 7.4 PBS, from 5 to 50 μM. (**B**) Calibration plot of the anodic peak current associated with the electroanalytical oxidation of DA over the concentration range for a bare SPE (black square), a 2.5% GO-SPE (orange circle), a 5% GO-SPE (blue triangle), a 7.5% GO-SPE (purple inverted triangle), and a 10% GO-SPE (green star). Error bars are on the data points and represent the average standard deviation (N = 3). Scan rate utilised: 100 mVs^{-1} (vs. SCE). (Figure reproduced from [21]. Creative Commons CC-BY license)

The use of carbon nano-onions have been made into a screen-printed ink and have shown to give rise to useful electroanalytical signal towards dopamine [22]. Fig. 1.10 shows the voltammetric response which has an improved signal over that of a commercially available screen-printed electrode, which is attributed to the presence of the additional oxygen-containing groups in the binder and is inferred to be a result of electrocatalytic reactions [22].

Other uses of screen-printed electrode can be highlighted by the use of molecularly imprinted polymer nanoparticles (nanoMIPs) in screen-printed electrodes for detecting cTnI in clinical patient serum samples post myocardial infarction using thermal detection [23]. MIPs are produced by self-assembly of specifically selected monomers and cross-linkers around the desired target, acting as a template. That said, MIPs have limitations which include heterogeneous binding affinity sites and slow binding kinetics. To overcome these drawbacks, nanoMIPs have been developed which are produced using a solid-phase approach, where peptide sequences derived from the target protein, cTnI, are covalently immobilised onto functionalized glass beads. Functional monomers are then polymerised around the immobilised template, forming nanoparticles that replicate the target's binding sites. A two-step elution process ensures that only the highest-affinity nanoMIPs are collected, while lower-affinity particles are removed.

As shown in Fig. 1.11, one can see the nanoMIPs-modified SPE is placed into a microfluidic cell where for each run, the nanoMIP-functionalized SPE was positioned at the centre of the copper block, with a rubber O-ring placed on top to prevent leakage. The resin cell was then assembled and secured with four nuts and

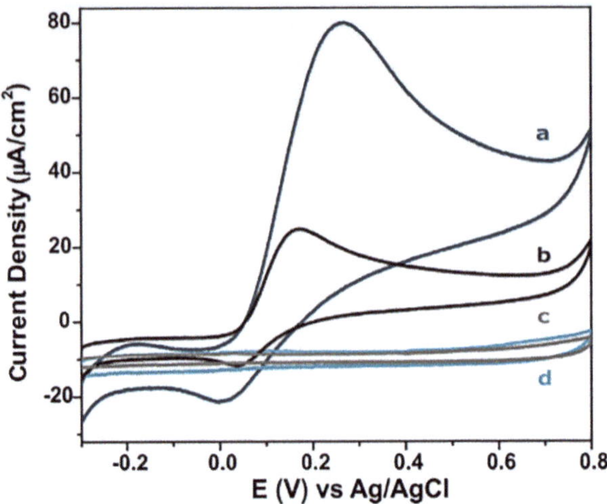

Fig. 1.10 Cyclic Voltammograms of (**a**) CNO/GRT SPE and (**b**) commercial SPE in the presence of 99.9 µM DA in PBS pH 7.4, (**c**) CNO/GRT SPE, and (**d**) commercial SPE in the absence of DA (only in PBS pH 7.4). Scan rate: 50 mV·s^{-1} (vs. pseudo Ag/AgCl). (Fig from ref. [22] Creative Commons CC-BY license)

bolts to ensure a tight seal before being placed in a controlled environment. Two type K thermocouples are placed in the measurement cell, with T1 positioned in the copper block heat sink and T2 located in the sample chamber, 4 mm above the functionalized SPE. The cell was then connected to the heat transfer device to monitor and quantify temperature changes. For all heat-transfer method measurements, three patient samples were tested per functionalized SPE. A typical experimental sequence consisted of: Blank—Sample 1—Sample 2—Sample 3—Blank, where the blank was human serum with no elevated cTnI levels (Fig. 1.11d). The first blank established the baseline thermal resistance (R_{th}) of the system, while the final blank confirmed that changes in R_{th} were due to cTnI binding rather than intrinsic serum properties. For each sample, 120 µL of serum or plasma was added to the sample chamber, and a copper lid was placed on top to minimise heat loss. The system was allowed to stabilise for 20 min; once a plateau was reached, 100 data points were recorded, and their mean and standard deviation calculated. After measurement, the sample was removed and replaced with the next, repeating the process for all samples. This study shows how one can provide a rapid and low-volume methodology to measure cTnI in clinical patient serum samples post myocardial infarction. Other approaches have made fabricated 2D hexagonal boron nitride for the oxygen reduction reactions [24], MnO_2 for hydrogen peroxide [25], Cobalt Phthalocyanine for citric acid [26] and glucose [27], WS_2 towards the hydrogen evolution reaction [28] and Prussian Blue for glucose sensing [29] to name just a few. These electrodes are widely used in electroanalysis and electrocatalytic applications. The use of screen-printing makes them cost-effective, reproducible, and suitable for portable sensing platforms, improves signal stability and reduces surface fouling compared to unmodified carbon-based electrodes.

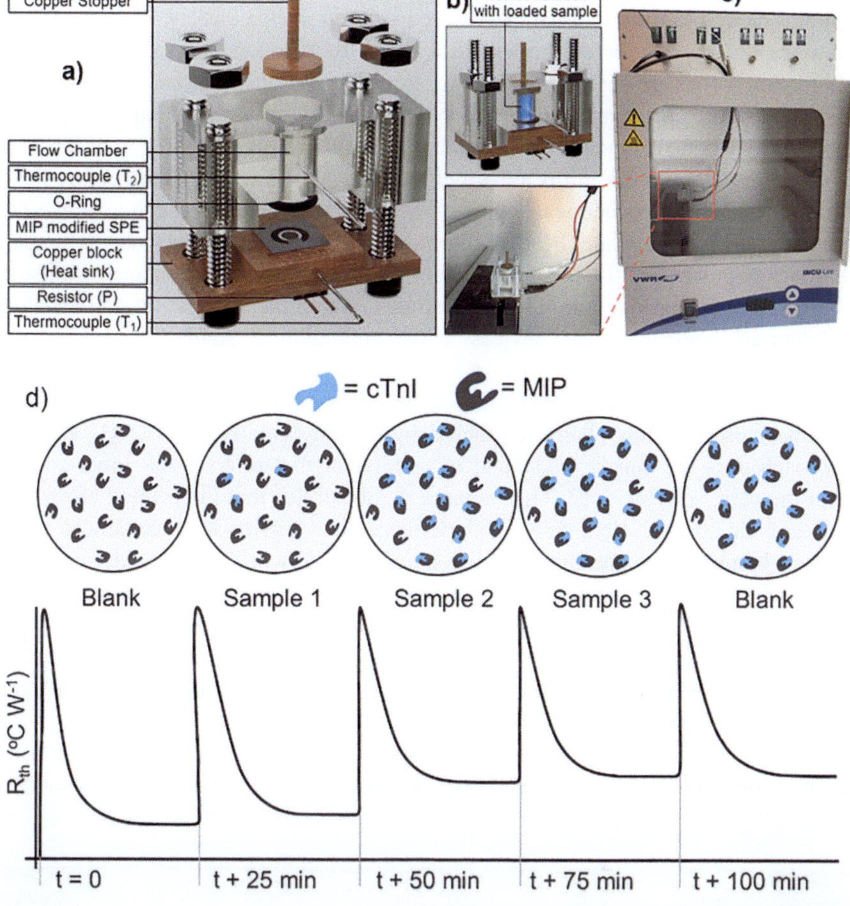

Fig. 1.11 Experimental set-up for cTnI concentration measurements with a focus on the (**a**) measurement cell design, (**b**) measurement cell under general operation, (**c**) final lab H heat-transfer method measurement setup with the measurement cell enclosed in a controlled environment (4250 × 2800 × 2600 mm), and (**d**) the typical plot generated for one set of HTM measurements with cTnI binding. (Figure from [23] Creative Commons CC-BY license)

1.4 Substrate

A key advantage of screen-printed electrodes is the remarkable flexibility in the choice of materials for constructing electrochemical cells. This versatility extends to both the substrate and the electrode layers, enabling the design of sensors tailored to specific analytical applications. Substrates play a critical role in defining the mechanical, chemical, and electrochemical properties of the sensors. Ceramic substrates are commonly employed due to their chemical inertness, thermal stability, and rigidity, making them suitable for high-temperature or harsh chemical environments. Paper-based substrates have also gained considerable attention because of their low cost, biodegradability, and capillary-driven fluid transport properties. Among paper platforms, researchers have explored a wide variety, including chromatographic papers for controlled fluid flow, filter papers for sample filtration, photo papers for high-resolution printing, stone papers for enhanced durability, vegetal papers derived from plant fibres for sustainability, waterproof papers for moisture resistance, and standard office papers for rapid prototyping.

In parallel, polymeric films are widely used as substrates for constructing flexible and robust electrochemical sensors. Thermoplastic materials such as polyester, polyethylene terephthalate (PET), polyimide films, and transparency sheets are particularly popular due to their mechanical flexibility, chemical resistance, and suitability for mass production. These substrates enable the fabrication of bendable, lightweight, and wearable devices while maintaining stable electrochemical performance.

Other notable work has used textile-based substrates where 'biosensors in briefs' have been used to sense chemicals such as hydrogen peroxide and NADH [30]. This approach uses the elastic waistband in underwear that offers tight direct contact with the skin, see Fig. 1.12. The research demonstrates that the printed electrodes maintain favourable electrochemical performance even under mechanical stress such as bending and stretching, which simulates real-world clothing deformation. Electrochemical tests with compounds like ferrocyanide, hydrogen peroxide, and NADH show reliable and reproducible results. The authors concluded that textile-based sensors are promising for future non-invasive monitoring in healthcare, sports, and military contexts, especially when tailored ink formulations and printing protocols are optimised for specific fabrics [30].

Other work has shown that wearable electrochemical sensors on underwater garments comprised of the synthetic rubber neoprene. The neoprene-based sensor was

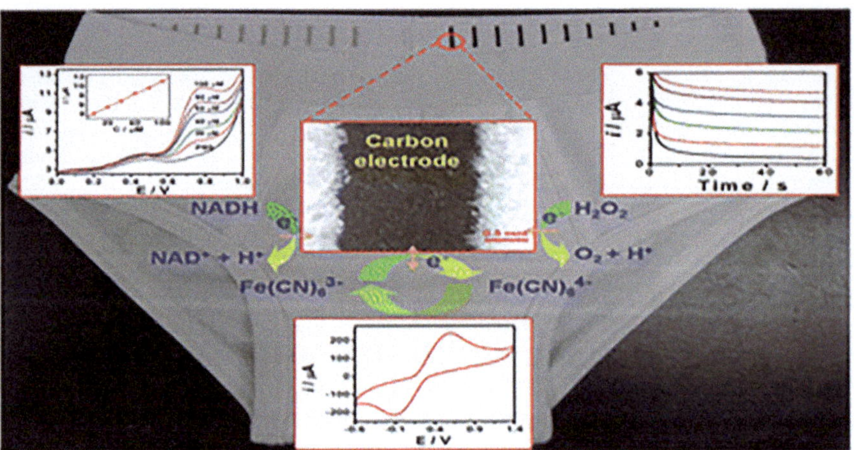

Fig. 1.12 Biosensors in briefs showing the measurement of ferrocyanide, hydrogen peroxide and NADH. (Reproduced from Ref. [30])

evaluated towards the voltammetric detection of trace heavy metal contaminants and nitro-aromatic explosives in seawater samples, with further applications involving the first example of enzyme (tyrosinase) immobilisation on a wearable substrate towards the amperometric biosensing of phenols [31]. This wearable sensing system offers a hands-free, continuous assessment tool for divers and surfers, with potential applications in environmental monitoring, public safety, and military operations.

Overall, the broad range of available substrates underscores the adaptability of screen-printed electrodes, supporting their use in diverse sensing applications from disposable point-of-care diagnostics to durable environmental monitoring devices. Generally, a good substrate is simply classed as it having a good adhesion of the screen-printing ink; the current diversity of substrates, as shown above, is fascinating.

1.5 Changing the Length of the Substrate

Changing the length of the substrate affects the performance of the screen-printed platform [32, 33]. It is interesting to note that the length of the SPE has remained constant in terms of the length of the connection length, typically 100 mm in length [34]. Whittingham et al [32] have made screen-printed electrodes with five different connection lengths of 32, 27, 22, 17, and 12 mm long and have characterised these towards a redox outer-sphere probe using cyclic voltammetry and electrochemical impedance spectroscopy and for the sensing of NADH and lead (II) ions (Fig. 1.13).

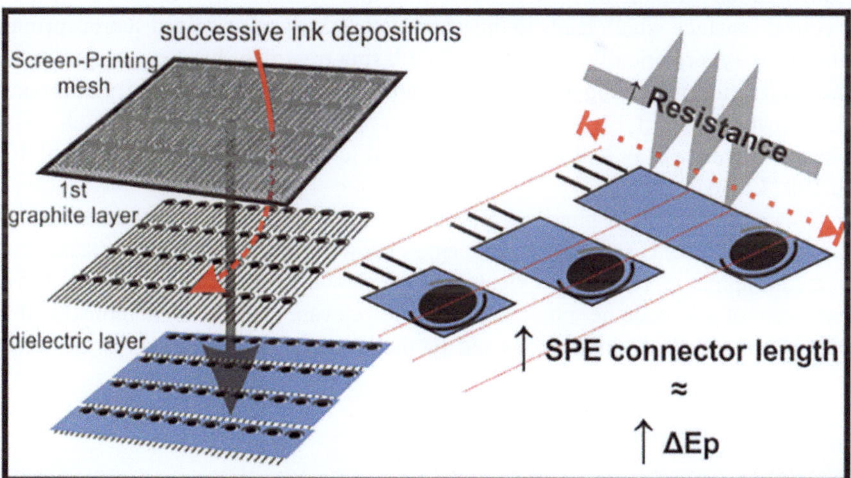

Fig. 1.13 An overview of changing the length of the SPEs reduces the resistance and give rise to improved electrochemical and electroanalytical responses. (Reproduced from Ref. [32])

The authors showed that the shorter electrode length can a lower resistance and improved electrochemical transfer rate constants and improved the sensing capabilities [32]. These findings are of high importance to those experimentalists working with screen-printed electrode sensors, and those who design their own screen -printed electrodes

1.6   Surface Activation

Chemical surface modification and electrochemical anodization have been used as effective strategies to improve the electrochemical activity of screen-printed electrodes. Chemical treatments typically introduce functional groups, catalytic sites, or nanostructured coatings that enhance electron transfer, whereas anodization provides a facile route to activate the electrode surface without the need for external modifiers. Electrochemical anodization has been used, for example in the case of the hydrogen evolution reaction [35]. This approach employs a constant potential, which is applied on the working electrode immersed in different electrolytes for a certain of time. The authors reported that sodium hydroxide shows the increased activity for the hydrogen evolution reaction. The authors states three key points: (1) a facile anodization treatment can improve the electrocatalytic activity of screen-printed electrode with apparently reduced overpotential and increased current density for the hydrogen evolution reaction.; (2) the activity enhancement of SPCEs is highly dependent on anodization conditions including electrolyte, treatment time, and anodization potential; (3) the introduction of negatively charged oxygen-containing functional groups during anodization, together with the increased defect sites and hydrophilicity on screen-printed electrode surface which leads to the promoted activity of anodized screen-printed platforms synergistically [35]. Recently, plasma treatment has been used for surface activation providing a reagent-free and environmentally friendly approach. As shown in Fig. 1.14, a hand-held atmospheric air plasma pen which has a plasma power of 3 W and increases the local temperature over the range of 100 to 190 °C [36].

This hand-held atmospheric air plasma pen is applied for 2 mins and one can see the changes in the C-O surface moieties, as shown in Fig. 1.15. These plasma pen-treated graphite screen-printed electrodes are shown to be useful for the sensing of paracetamol and codeine, pharmacologically relevant analgesics, demonstrated in both pharmaceutical formulations and synthetic saliva successfully.

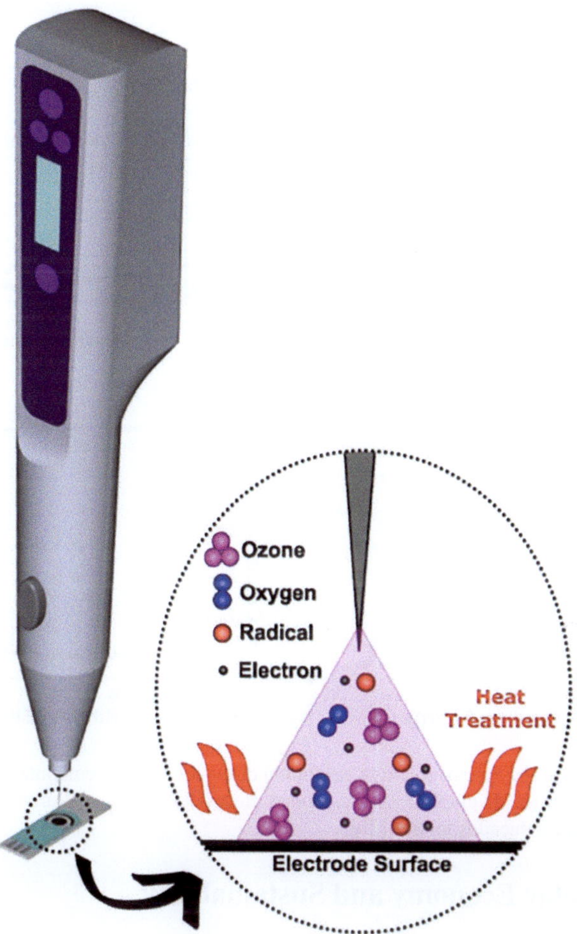

Fig. 1.14 Schematic representation of the plasma pen and the illustration of the plasma effect on SPE surface. (Reproduced from Ref [36])

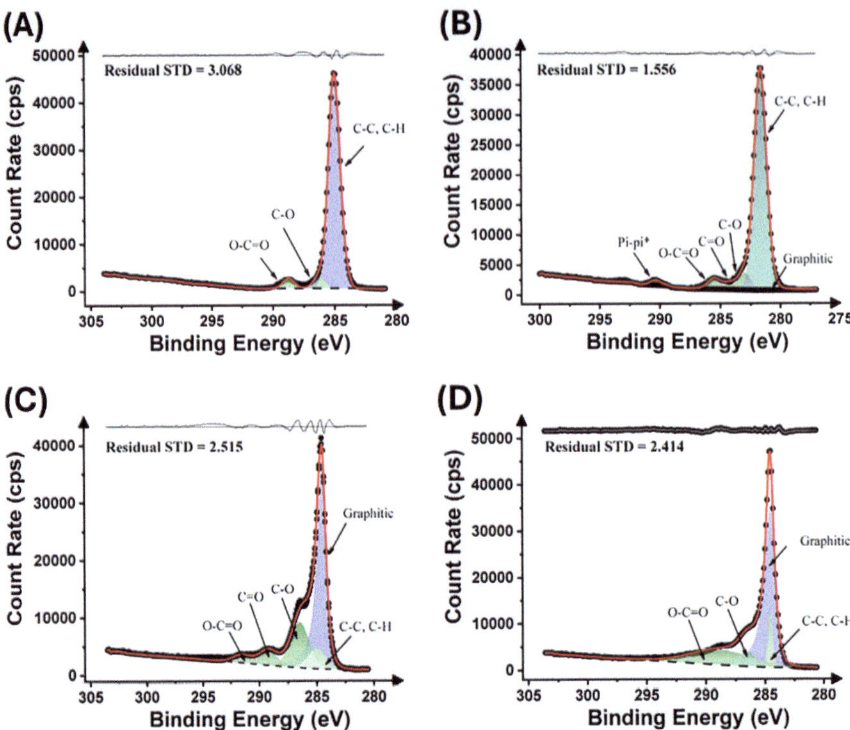

Fig. 1.15 XPS survey spectra and high-resolution deconvoluted C1s components obtained for (**a, b**) BDD-SPE and (**c, d**) graphite screen-printed electrodes before and after plasma treatment, respectively. (Reproduced from Ref [36])

1.7 Circular Economy and Sustainability

Adopting circular economy principles involves designing devices and fabrication processes that minimise waste, extend sensor lifetimes, and enable reuse or recycling of components. This includes developing electrode platforms that can be regenerated through simple cleaning or surface activation, designing modular sensor systems with replaceable sensing elements, and using materials that can be recovered or repurposed at end-of-life. Sustainability in this context refers to the responsible selection of materials, manufacturing methods, and disposal strategies that reduce environmental and health impacts. Sustainable approaches emphasise the use of abundant, low-toxicity, or bio-derived materials (e.g., cellulose substrates, biodegradable polymers, carbon-based nanomaterials), as well as green synthesis techniques for inks and modifiers. Integrating circular economy and sustainability concepts into electrochemical sensor design not only lowers the ecological footprint but also improves economic feasibility and aligns with regulatory and societal pressures for greener analytical technologies. Within screen-printed electrode platforms, several strategies have been reported: (1)

regeneration and reuse: screen-printed electrodes can be reactivated through simple treatments such as electrochemical cleaning, plasma treatment, or anodization, allowing their repeated use without the need for entirely new electrodes; (2) *biodegradable substrates:* screen-printed electrode fabricated on recycled paper, cellulose, or bioplastic films reduce reliance on petroleum-based plastics and facilitate eco-friendly disposal after single use; (3) *green conductive inks*: development of carbon-based inks derived from biomass (e.g., lignin, biochar) or waste-derived nanomaterials provides sustainable alternatives to traditional graphite or metal ink; (4) *metal recovery*: screen-printed electrochemical platform modified with precious metals (e.g., Au, Pt, Ag) can be processed at end-of-life to recover these valuable materials, fitting within circular material loops. Sustainability in SPE-based sensors is therefore strongly linked to material choice, ease of regeneration, and strategies for end-of-life recovery. Together, these approaches lower the ecological footprint of disposable sensors while enabling scalable and cost-effective production.

For example, Cumba and co-workers [37] have designed and prepared a novel, environmentally friendly screen-printable ink formulation for Ag/AgCl reference electrodes, designed to reduce the environmental and health risks associated with traditional silver nanoparticle-rich inks. The new ink contains only 25% silver nanoparticles, uses a non-toxic solvent, and incorporates a biodegradable polymer binder. These electrodes demonstrate excellent electrochemical stability, low resistance, and high reproducibility in glucose sensing applications. Cytotoxicity tests show significantly lower toxicity compared to commercial inks, making them suitable for wearable and implantable biosensors. The electrodes also maintain performance over time, with minimal degradation after 6 months, and show strong correlation with commercial glucose test strips in real blood sample analysis. This innovation offers a sustainable alternative for disposable biosensors, particularly in diabetes monitoring [37]. Another approach has reported a sustainable approach to fabricating disposable screen-printed electrodes using biochar derived from peanut shells as a carbon source [38]. As shown in Fig. 1.16, peanut shells, an abundant agricultural by-product were repurposed into biochar through a low-cost pyrolysis process using a home-built kiln.

The peanut shell biochar was chemically activated using organic solvents, alkaline, and acidic treatments to enhance its surface properties, porosity, and electrochemical performance. Once processed, the biochar was mixed with nail polish and acetone to create a printable ink, which was then applied to PET substrates to fabricate screen-printed electrodes. These biochar-based electrodes were successfully used to detect paracetamol in aqueous solutions, demonstrating both environmental sustainability and practical analytical performance.

The applicability of circular economy strategies for screen-printed electrodes is strongly influenced by the matrix in which detection occurs. For example, in relatively clean aqueous samples, electrodes may be regenerated and reused multiple times through electrochemical cleaning or surface activation. However, in complex biological matrices such as blood, saliva, or urine, electrode fouling becomes a major limitation and these need to be disposed of appropriately.

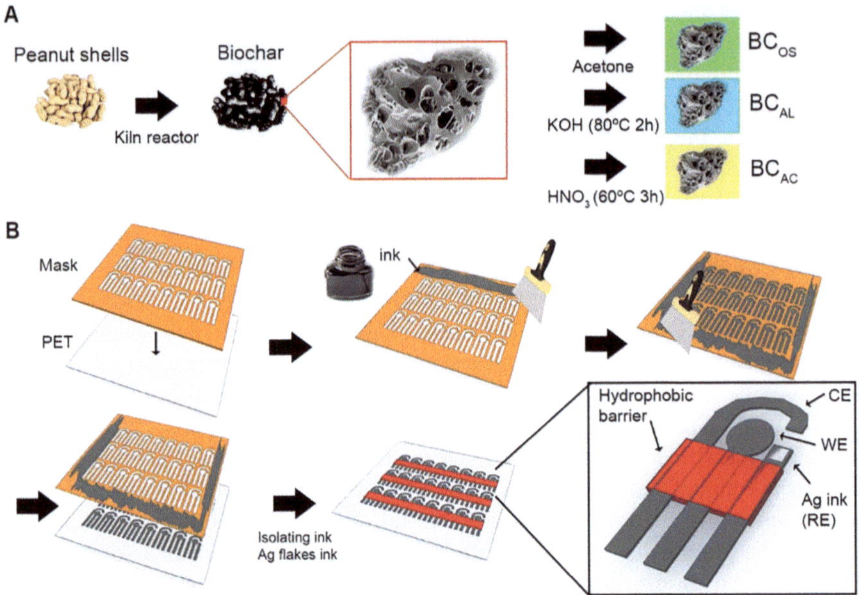

Fig. 1.16 Scheme showing in brief the process to obtain pristine and activated biochar (**A**). Methodological procedure to obtain the screen-printed electrodes (**B**). (Reproduced from Ref [38] Creative Commons CC-BY license)

References

1. A. García-Miranda Ferrari, S.J. Rowley-Neale, C.E. Banks, Talanta Open **3**, 100032 (2021)
2. R.D. Crapnell, C.E. Banks, ChemElectroChem **11**, e202400370 (2024)
3. S. Singh, J. Wang, S. Cinti, ECS Sensors Plus **1**, 023401 (2022)
4. J.P. Metters, R.O. Kadara, C.E. Banks, Analyst **136**, 1067–1076 (2011)
5. K.K. Mistry, K. Layek, A. Mahapatra, C. RoyChaudhuri, H. Saha, Analyst **139**, 2289–2311 (2014)
6. A. Heller, B. Feldman, Chem. Rev. **108**, 2482–2505 (2008)
7. S. Eissa, N. Alshehri, A.M.A. Rahman, M. Dasouki, K.M. Abu-Salah, M. Zourob, Biosens. Bioelectron. **101**, 282–289 (2018)
8. J. Chouler, Á. Cruz-Izquierdo, S. Rengaraj, J.L. Scott, M. Di Lorenzo, Biosens. Bioelectron. **102**, 49–56 (2018)
9. Y. Xu, M.G. Schwab, A.J. Strudwick, I. Hennig, X. Feng, Z. Wu, K. Müllen, Adv. Energy Mater. **3**, 1035–1040 (2013)
10. L.O. Orzari, L.R.G.e. Silva, R.C. de Freitas, L.C. Brazaca, B.C. Janegitz, Microchim. Acta **191**, 76 (2024)
11. J. Wang, B. Tian, V.B. Nascimento, L. Angnes, Electrochim. Acta **43**, 3459–3465 (1998)
12. J.R. Camargo, R.D. Crapnell, E. Bernalte, B.C. Janegitz, C.E. Banks, ACS Appl Electr Mat **7**, 5599–5610 (2025)
13. J.P. Metters, F. Tan, R.O. Kadara, C.E. Banks, Anal. Methods **4**, 1272–1277 (2012)
14. J.P. Metters, R.O. Kadara, C.E. Banks, Analyst **137**, 896–902 (2012)
15. J.L. Chang, J.M. Zen, Electroanalysis **18**, 941–946 (2006)
16. L. Authier, C. Grossiord, P. Brossier, B. Limoges, Anal. Chem. **73**, 4450–4456 (2001)

17. J.-L. Chang, J.-M. Zen, Electrochem. Commun. **8**, 571–576 (2006)
18. J.P. Metters, R.O. Kadara, C.E. Banks, Analyst **138**, 2516–2521 (2013)
19. P. Roberto de Oliveira, R.D. Crapnell, A. Garcia-Miranda Ferrari, P. Wuamprakhon, N.J. Hurst, N.C. Dempsey-Hibbert, M. Sawangphruk, B.C. Janegitz, C.E. Banks, Biosens. Bioelectron. **228**, 115220 (2023)
20. R.O. Kadara, N. Jenkinson, C.E. Banks, Electroanalysis **21**, 2410–2414 (2009)
21. S.J. Rowley-Neale, D.A.C. Brownson, G. Smith, C.E. Banks, Biosensors **10**, 27 (2020)
22. L.R. Cumba, A. Camisasca, S. Giordani, R.J. Forster, Molecules **25**, 3884 (2020)
23. J. Saczek, O. Jamieson, J. McClements, A. Dann, R.E. Johnson, A.D. Stokes, R.D. Crapnell, C.E. Banks, F. Canfarotta, I. Spyridopoulos, A. Thomson, A. Zaman, K. Novakovic, M. Peeters, Biosens. Bioelectron. **282**, 117467 (2025)
24. A.F. Khan, A.G.-M. Ferrari, J.P. Hughes, G.C. Smith, C.E. Banks, S.J. Rowley-Neale, Sensors **22**, 3330 (2022)
25. J. Zbiljić, V. Guzsvány, O. Vajdle, B. Prlina, J. Agbaba, B. Dalmacija, Z. Kónya, K. Kalcher, J. Electroanal. Chem. **755**, 77–86 (2015)
26. K.C. Honeychurch, L. Gilbert, J.P. Hart, Anal. Bioanal. Chem. **396**, 3103–3111 (2010)
27. E. Crouch, D.C. Cowell, S. Hoskins, R.W. Pittson, J.P. Hart, Anal. Biochem. **347**, 17–23 (2005)
28. J.P. Hughes, F.D. Blanco, C.E. Banks, S.J. Rowley-Neale, RSC Adv. **9**, 25003–25011 (2019)
29. F. Ricci, A. Amine, C.S. Tuta, A.A. Ciucu, F. Lucarelli, G. Palleschi, D. Moscone, Anal. Chim. Acta **485**, 111–120 (2003)
30. Y.-L. Yang, M.-C. Chuang, S.-L. Lou, J. Wang, Analyst **135**, 1230–1234 (2010)
31. K. Malzahn, J.R. Windmiller, G. Valdés-Ramírez, M.J. Schöning, J. Wang, Analyst **136**, 2912–2917 (2011)
32. M.J. Whittingham, N.J. Hurst, R.D. Crapnell, A. Garcia-Miranda Ferrari, E. Blanco, T.J. Davies, C.E. Banks, Anal. Chem. **93**, 16481–16488 (2021)
33. P. Wuamprakhon, A.G.-M. Ferrari, R.D. Crapnell, J.L. Pimlott, S.J. Rowley-Neale, T.J. Davies, M. Sawangphruk, C.E. Banks, Sensors **23**, 1360 (2023)
34. S.A. Wring, J.P. Hart, L. Bracey, B.J. Birch, Anal. Chim. Acta **231**, 203–212 (1990)
35. X. Niu, L. Shi, X. Li, J. Pan, R. Gu, H. Zhao, F. Qiu, Y. Yan, M. Lan, Electrochim. Acta **235**, 64–71 (2017)
36. M. Di-Oliveira, R.G. Rocha, M.C. Marra, T.C. Oliveira, M. Vojs, M. Marton, R.D. Crapnell, C.E. Banks, E.M. Richter, R.A.A. Muñoz, Electrochim. Acta **537**, 146851 (2025)
37. L.R. Cumba, R. Byrne, F.Ó. Maolmhuaidh, A. Morrin, R.J. Forster, Electrochim. Acta **501**, 144797 (2024)
38. F. Bernassani, M. Mosquera-Ortega, I. Sánchez, S. Susmel, E. Cortón, F. Figueredo, Anal. Methods **17**, 6565–6576 (2025)

Chapter 2
Fundamentals of Screen-Printing Electrochemical Architectures

Contents

2.1 Screen-Printing Process

Screen-printing is widely recognised for producing thick-film hybrids, with applications ranging from circuit boards to electrochemical devices. Its key advantage lies in enabling mass production of thick films through low-cost, straightforward designs, offering strong economic scalability. Owing to the simplicity of the equipment, the process can also be adapted and optimised once the fundamental principles of screen-printing are understood. To achieve consistent and reproducible thick films, five essential elements are required: (1) suitable printing medium; (2) mesh screen with an embedded stencil design; (3) substrate for deposition; (4) a flexible, durable squeegee; and (5) a secure base to stabilise the substrate during printing.

While appropriate machinery is important for ensuring reproducibility and high throughput, in some cases, the equipment itself can hinder print quality. Thus, successful screen-printing ultimately depends on a well-controlled methodology.

Upon integration of the above-mentioned prerequisites, a print cycle can occur; Fig. 2.1 represents the process in three simple steps. The first step consists of placement of the printing media (i.e., ink, emulsion or paste) upon the mesh screen,

C. E. Banks, R. D. Crapnell, *Screen-Printing Electrochemical Architectures*, SpringerBriefs in Applied Sciences and Technology, https://doi.org/10.1007/978-3-032-10411-3_2

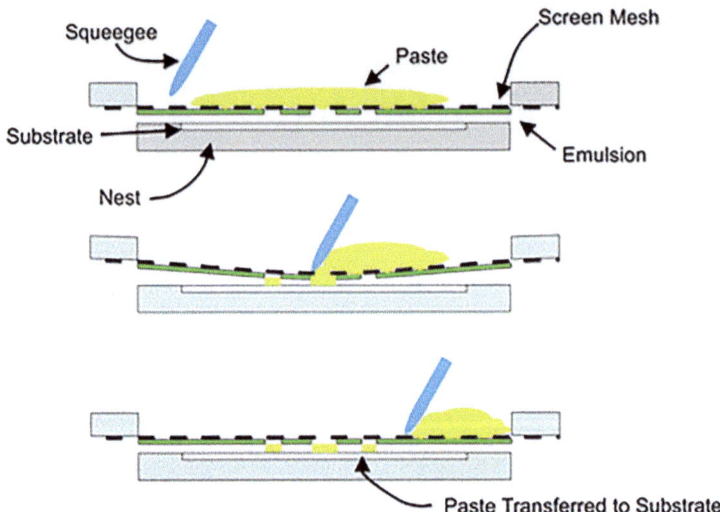

Fig. 2.1 Schematic of the printing process. (Figure from [1])

where it is visible that the screen is not in contact with the substrate. Such contact could potentially damage or spoil the print as the substrate would uncontrollably 'snap-off' the screen, creating an unclean print. The contact only occurs when the squeegee applies pressure over the screen forcing the printing medium through the stencil design, creating the desired pattern or design in a controlled and efficient manner.

2.2 Selection of Screen-Printing Equipment

As mentioned earlier, the prerequisites are essential for the screen-printing process. Equally important is the careful selection of equipment, as this plays a crucial role in maximising reproducibility and output. This section therefore considers the screen, squeegee, flood/distribution bar, and printing media.

2.2.1 The Screen

When designing and using a screen, three main factors must be considered: the screen frame, stencil design, and screen-mesh. The screen frame is typically fabricated from metal, as the material must withstand the high tension exerted by the mesh, which in some cases can exceed 80+ kg. While the frame size may vary, screens are generally either rectangular or square, depending on the machine and design requirements. Consequently, there is no universal ratio or dimension for an optimal design; instead, reference is made to the internal dimensions of the screen, since the frame functions solely as structural support for the open mesh area (calculated using Eq. 2.1). In practice, two primary frame types are used in screen-printing: lightweight cast aluminium and extruded aluminium. The appropriate choice depends on the specific machine and application.

$$\text{Open Area} = \frac{\left(\text{mesh area}\right)^2}{\left(\text{wire diameter mesh opening}\right)^2} \times 100\% \tag{2.1}$$

The screen-mesh possesses several key properties that enable the production of high-quality screen-printed designs. It functions as a support network, holding the stencil design securely in place throughout the printing cycle, even under the pressure applied by the squeegee. These properties are determined by both the design and manufacturing of the mesh, making material selection critical for achieving precise and detailed prints. Figure 2.2 presents an optical image of a screen-mesh, highlighting its woven structure. Within this mesh, the electrodes will be fabricated. The spacing of the mesh is particularly important, as it governs the amount of ink that can pass through and be deposited onto the substrate.

A factor that must be considered upon creation of the mesh screen is the mesh count, M, i.e., the number of wires per unit of length as shown in Fig. 2.3. In conjunction with size of the mesh count and wire diameter, D, the mesh opening can be calculated via Eq. 2.2. The mesh opening is a vital measurement that influences the amount and size of the ink that can be passed through the screen.

$$O = \frac{1}{M} - D \tag{2.2}$$

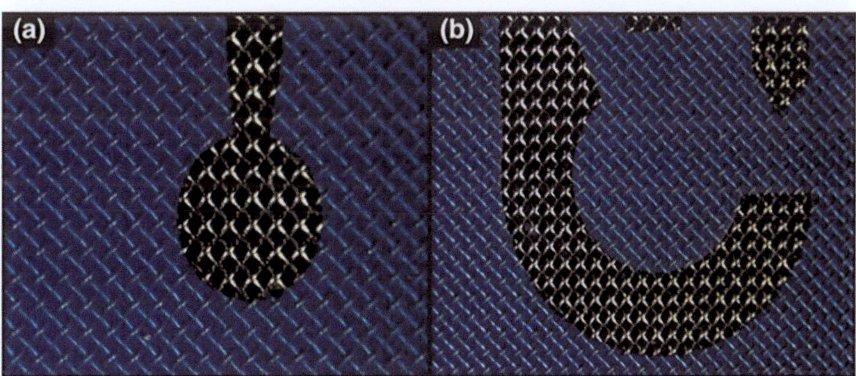

Fig. 2.2 Optical image of a stainless steel screen-mesh, for a working electrode (*left image*) and counter electrode (*right image*)

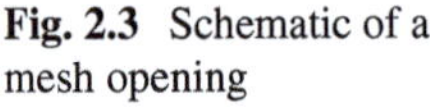

Fig. 2.3 Schematic of a mesh opening

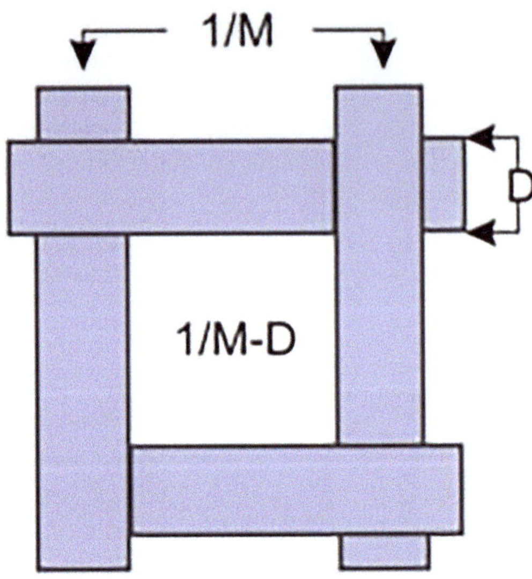

Table 2.1 Representation of the capabilities of the three materials for the design of mesh screens

Screen requirements	Polyester	Stainless steel	Nylon
Flexibility	1	2	3
Resilience	1	2	3
Percentage of open area	2	1	2
Stability of print size	2	1	3
Damage from squeegee	2	3	1
Accidental damage	2	3	1
Cost	1	3	1
Minimal snap-off from large areas	2	1	3

Ranked 1–3 which corresponds to the worst (3) and best (1) in its category

Many manufacturers prefer to use three types of materials; these being nylon, polyester and stainless steel. Presented in Table 2.1 are the benefits of utilising each material. Along with careful consideration of Table 2.1 other vital observations must be carried out. For example, the minimum line width that can be printed must be three times the mesh thread diameter, for example, large mesh threads cannot print small designs. In addition, the mesh diameter must also be three times larger than the particulate size of the printing medium, to allow suitable passage of the media, this is especially important when considering conductive inks. Furthermore, to the above-mentioned mesh designs, it is possible to utilise a V-mesh, which consists of a mesh created from Vecry which is a sheathed filament surrounding a liquid crystal-based polymer core.

This mesh provides the required strength for a screen, the combination of thin, flexible fibres and the weaving process result in flat filament intersections. Additionally, the smoother thread surfaces further enhance paste release. Such

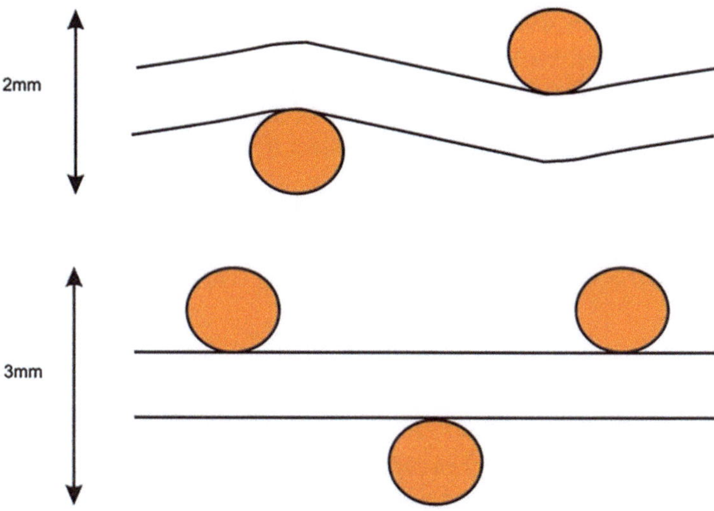

Fig. 2.4 Representation of the effect on heat rolling of polyester and nylon mesh screens

screens can exhibit higher levels of tension allowing less-snap off, thus reducing the risk of faulty screen-prints. The thickness of the screen is additionally vital for the printing process, in most cases, the overall screen-mesh thickness will be double that of the wire diameter. Typically, within the fabrication process of the polyester and nylon mesh screens the mesh is woven and then passed through heat rollers which weld the wires together creating a bound matrix of mesh wires. In the case of stainless steel meshes the metallic properties of the steel allows for instant mesh strength upon weaving of the wires, therefore this additional heating step is no longer required. Represented within Fig. 2.4 are the resulting wire mesh thicknesses upon utilisation of the two fabrication methods. It is noticeable that a distortion of the wire mesh occurs upon utilisation of the heat rollers, therefore reducing the wire diameter from 3 to 2 mm. It is important to note that the stainless steel mesh thicknesses can vary an extraordinary amount due to the nature of the stiffness and hardness of the steel wires.

As mentioned previously, the amount of printing medium that can be transferred relies on the opening in the screen and the thickness, the volume of paste (V) can be calculated using Eq. 2.3. Assuming the screen thickness is double that of the wire diameter:

$$V = \left(2D\right)\left(\frac{1}{M} - D\right)^2 \tag{2.3}$$

In order for a successful print with minimal snap-off the screen-mesh must be held at a specific tension, which must be sufficient to allow the screen to peel away from the substrate in a controlled and reproducible fashion. Nonetheless excessive tension upon the mesh can cause detrimental and costly damage. It is with manipulation of these tensile qualities that a perfect mesh can be created. It is noted that screens that are highly taught will give the best control of the snap-off and therefore reproducibility of the print. On the other hand, highly taught screens are more likely to be susceptible to unwanted or accidental damage due to their excessive tension. Therefore, when utilising stainless steel screens, the elongation of 0.9 % will allow the screen to revert back to its original state, with an additional 0.1 % as leverage. It is possible for a mesh screen to give a perfectly defined and reproducible print at values of half that of the elongation limit, considering that an alteration to the print gap between the screen and substrate is incurred, for the perfect snap-off.

2.2.2 The Squeegee

A critical factor during the screen-printing cycle is the consistent movement of the printing medium towards the substrate at an appropriate speed and pressure, achieved using a squeegee. As outlined in Sect. 2.1, the squeegee applies controlled pressure to force the ink through the stencil design, producing a uniform print on the substrate. For the squeegee to perform effectively, it must be fabricated from materials that combine flexibility with resilience. Polyurethane is the most widely used material in industry, as it provides durability; under typical conditions, a squeegee can achieve up to 20,000 prints before significant wear is observed. However, when stainless steel meshes are employed, this lifetime is reduced due to increased surface friction, which accelerates wear at the point of contact. Different grades of softness are available during fabrication: softer grades produce a larger contact area with the screen, thereby improving transfer efficiency. Note that the detrimental effect of squeegee wear on ink coverage results incomplete snap-off results in irregular and inconsistent prints. In practice, the squeegee should be at least 10 mm wider than the intended print area. Larger squeegees increase the applied pressure on the screen-mesh, reducing the natural print gap and potentially causing irreversible mesh damage (Fig. 2.5). Another critical parameter is the angle at which the squeegee applies pressure. At excessively steep angles, insufficient printing medium is transferred, resulting in a thinner ink or paste layer on the substrate. Conversely, at shallow angles, the hydrodynamic pressure increases, depositing too much

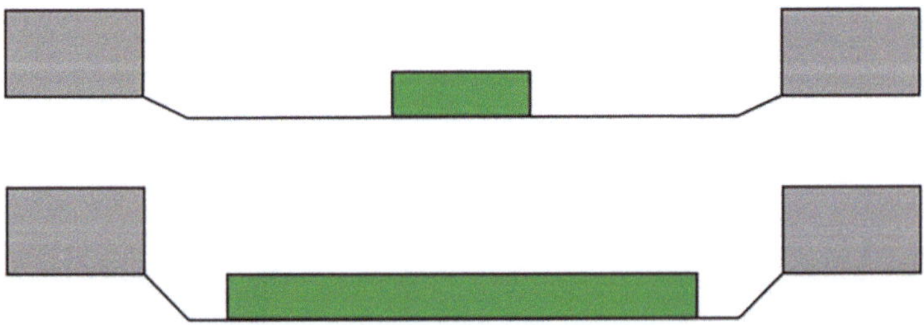

Fig. 2.5 Representation of the squeegee size and its effect upon the screen-mesh, where it can be seen that upon utilisation of a larger/heavier squeegee a naturally smaller print gap is created

medium and potentially blocking the stencil or hindering the snap-off process. Figure 2.6 presents the examples of different squeegee angles, highlighting their advantages and limitations within the screen-printing process.

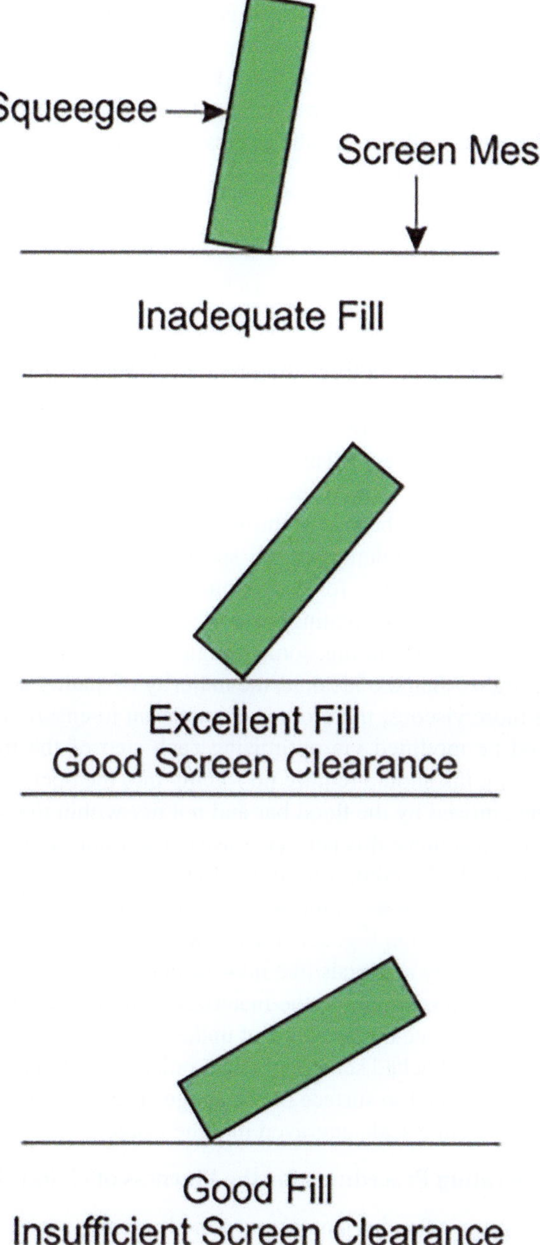

Fig. 2.6 Angles of the squeegee and their effect upon the screen-print

2.2.3 The Flood Bar/Distributor

The distribution of ink/paste is another vital process of the print cycle for a successful print. Typically flood bars are designed from stainless steel and will be slightly longer than the squeegee being utilised, it is then fixed behind the squeegee and upon the first transition of the print cycle the flood bar will transport the ink/paste over the print area. It is vital that the flood bar is slightly above or touching the screen-mesh, so that it is constantly in touching distance of the ink. In many printing situations the squeegee will remove the excess ink/paste back to the starting position, and the process can be repeated.

2.2.4 Printing Medium

The screen-printing process allows the user to fabricate a variety of geometric designs and shapes; such ability requires the utilisation of a durable, compatible printing medium. Terminology of such media can range from a dye to a simple ink or paste, but in most cases, they will all possess the same composition. The viscosity of these 'inks' will determine how successful the print is, as due to the nature of screen-printing, the ink must be passed through a specific shape (stencil) keeping its geometric design. The formulation of the specialist inks tends to consist of two components: suitable pigments for the application at hand and an appropriate amount of solvent/binder ratio creating the perfect transport of the pigment.

Upon application of screen-printing, some designs or screen-prints may require viscosities that are higher than the ideal, as the majority of quality control issues arise when the ink is more viscous, therefore it is important to ensure that the operation procedure should be modified via optimising each step of the manufacture. The ideal screen-printing ink should require no forcing into the open area of the screen, flow readily when moved by the flood bar and not dry within the screen-mesh during the operation. To achieve this perfect consistency, prior mixing of the ink must occur until a smooth fluidic composition is achieved.

In general examination of the composition is performed by utilising a fineness of grind (FOG) gauge (shown in Fig. 2.7) for the determination of dispersion, particle size, and fineness of many materials like inks, lacquers, pigments, filler, and chocolate to name a few. In this situation, the materials being tested are inks. The gauge may also be used to indicate the presence of undesired large particles in these materials. Such tool has an attached scrapper, which pulls the material along the sloping groove machined onto the top surface of the gauge. The value for fineness of grind is obtained directly from a scale engraved into the gauge.

Standard Operating Procedures for the Fineness of Grind Gauge

1. Place the gauge on a flat, horizontal and non-slip surface, with the zero mark on the scale closet to the user.
2. Place a considerable amount of material (ink) in the deep end of each groove.

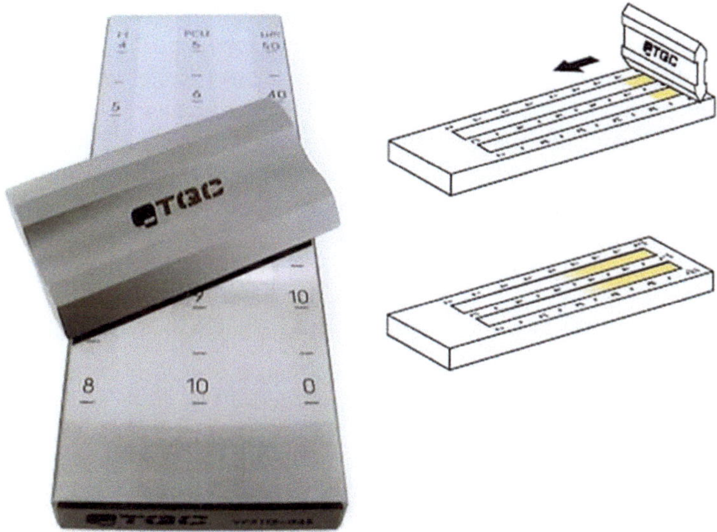

Fig. 2.7 Optical images of fineness of grind gauge

3. Place the scrapper on the surface of the gauge behind the material (ink), which is at the deepest groove.
4. Use both hands to hold the scrapper and pull along the length of the gauge at a constant speed apply sufficient pressure to clean the excess material (ink) from the edge of the gauge.
5. Stop at a point beyond the zero depth and assess the drawn-out material within the next 3 s.
6. Note: This avoids inaccurate testing due to evaporation of solvents from materials.
7. The material (ink) should be viewed at right angles to the length of the groove and at an angle of 20–30 with the surface of the gauge.
8. Find the first position across the groove 3 mm wide which contains 3–10 particles/streaks/scratches of material (ink).
9. Read the position on the scale and record this value.
10. Perform the test three times; afterwards calculate the average value of the result. The average value is the fineness of grind of the material (ink).

The FOG test can be used as a QA/QC measurement for inks used in the screen-printing process.

2.2.5   Curing of the Ink

The nature of the printing media requires a curing step within the fabrication of the design. Generally, many failures in the fabrication of conductive inks are reliant upon the poor selection of solvents. It is with this in mind that consideration of the solvents utilised within the 'ink' allow for an ideal curing time and temperature to envisage a situation where it is clear that upon the curing step of the ink, the volume starts to decrease, leaving behind a fully conductive 'stack' upon the substrate where the polymer in the ink formulation holds the structure of the electrode surface. In many cases, the manufacturer of the ink will provide the solvents that are the best ratio for working and curing properties, therefore if an amendment is requested consideration of the curing procedure must be endured. Note that an ink has a range of solvents varying from slow and fast evaporation time, to ensure a controlleddrying process resulting in a reproducible electrode surface. It is important to note that the solvent is duly there to create a fluidic support for the conductive paste and thus in most cases a solvent with a lower boiling point would be ideal.

Reference

1. J.P. Metters, R.O. Kadara, C.E. Banks, Analyst **136**, 1067–1076 (2011)

Chapter 3
Fabricating Screen-Printed Electrochemical Architectures: Successful Design and Fabrication

Contents

3.1 A Practical Guide to Screen-Printing

3.1.1 Standard Operating Procedure for Screen-Printing

In consultation with the prior section upon choosing the appropriate screen mesh, design, squeegee, and printing medium, the fabrication of the screen-printed design can begin. Within this section, a precise standard operating procedure for the manufacture of screen-printed electrodes is now described.

1. It is important to note that screen-printing is a completely clean process and as such, full protective clothing must be utilised. Therefore, a full body cleanroom suit, chemical-resistant plastic gloves and hairnet are a necessity, typically in a clean room environment. All exposed skin, including the neck, wrists, and arms, must be fully covered to prevent contamination and contact with printing inks. Typically, sleeves of the cleanroom suit should be tucked into gloves, and additional protective layers such as disposable arm sleeves may be used if necessary.

© The Author(s), under exclusive license to Springer Nature
Switzerland AG 2026
C. E. Banks, R. D. Crapnell, *Screen-Printing Electrochemical Architectures*,
SpringerBriefs in Applied Sciences and Technology,
https://doi.org/10.1007/978-3-032-10411-3_3

2. Select the appropriate screen design: Choose the screen design that corresponds to the desired electrode system. Screen-printed electrodes typically consist of three distinct ink layers; therefore, ensure that the correct screen and corresponding ink layer are selected for each stage of printing.

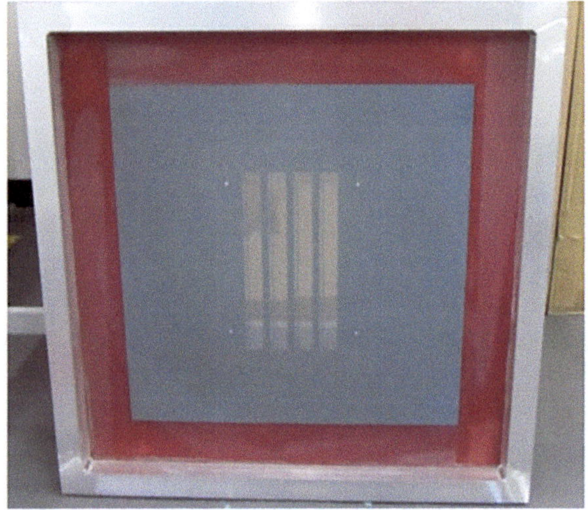

3. Pick up the screen holding the frame (not the mesh). As so not to damage or contaminate it.

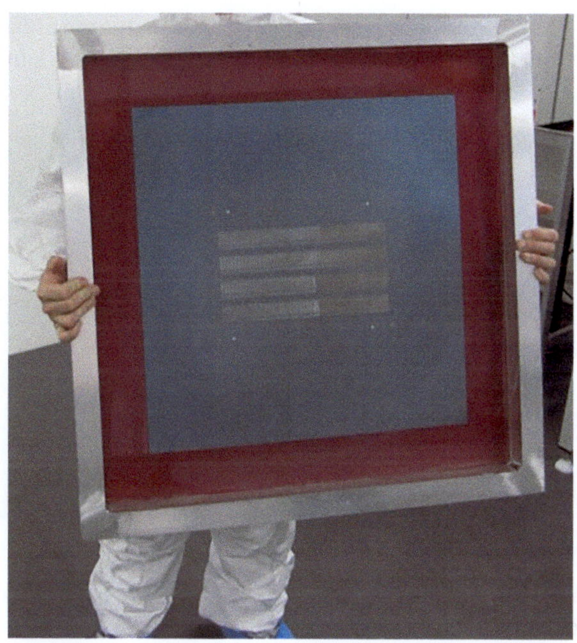

4. To guarantee the successful transfer of ink to the substrate the tension of the screen must be measured, to determine if the perfect snap-off will occur. Note: the value for this depends on the starting tension measurement and screen mesh specification. This parameter can also be used to inform you when a new screen is required after a long use.

5. Ensuring that the screen is the right way round (with the top of the screen facing downwards), carefully place and secure the screen into the printing machine.

6. After securing the screen into the machine, the printing medium can now be prepared. In general, due to the viscosity of the printing paste/ink extensive mixing of the ink must occur to create a fluidic homogenous mixture. This can be either done vigorously by hand or mechanically.

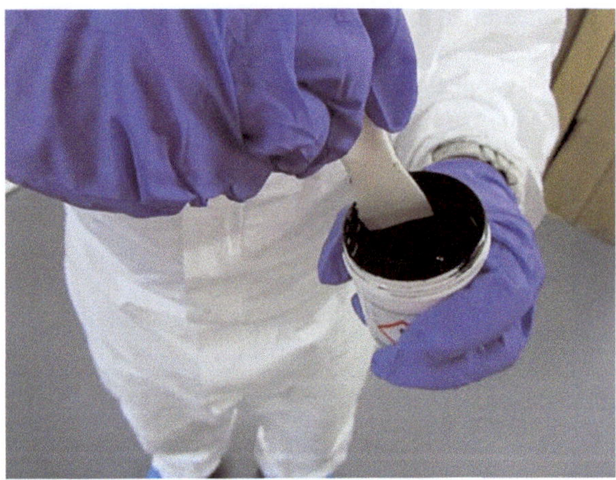

7. Once the viscosity of the paste is reduced, it is now ready for application upon the screen within the machine. It is vital that the ink is applied next to the screen design, with a suitable amount to print the desired quantity of electrodes, thus reducing the potential waste and costs.

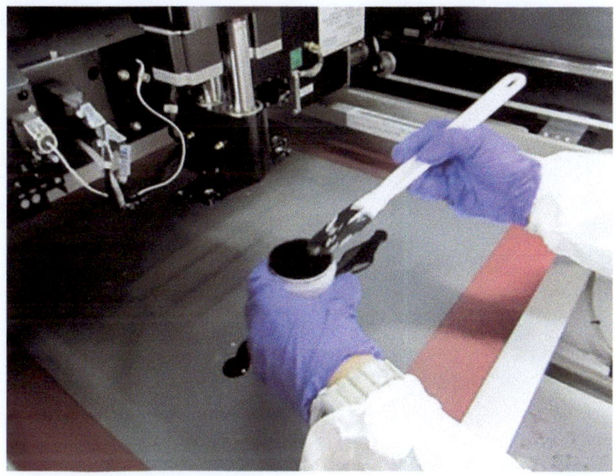

8. Now that the printing medium has been applied, the placement of the squeegee and flood bar can proceed. Firstly, place the flood bar on the supporting printer head, ensuring that the bolts are securely fastened. Secondly, place the clean squeegee in front of the flood bar and secure the bolts in a similar fashion to the first step. It is important to note that to stop contamination from different inks/pastes the utilisation of different flood bars and squeegees is considered best practice for each ink used.

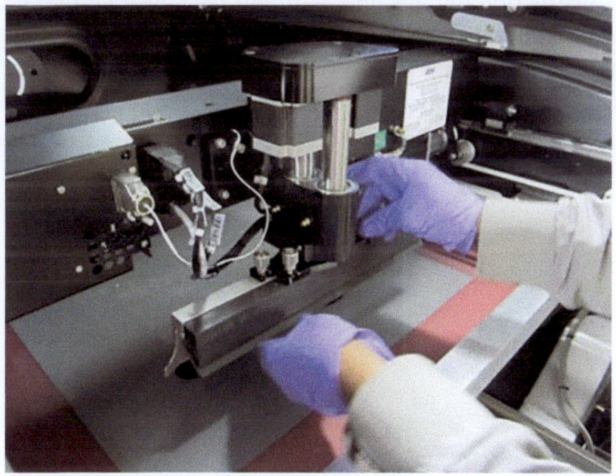

9. Placement of the desired substrate into the machine can now occur. It is extremely necessary that you are as reproducible as possible, so that each layer overlays successfully.

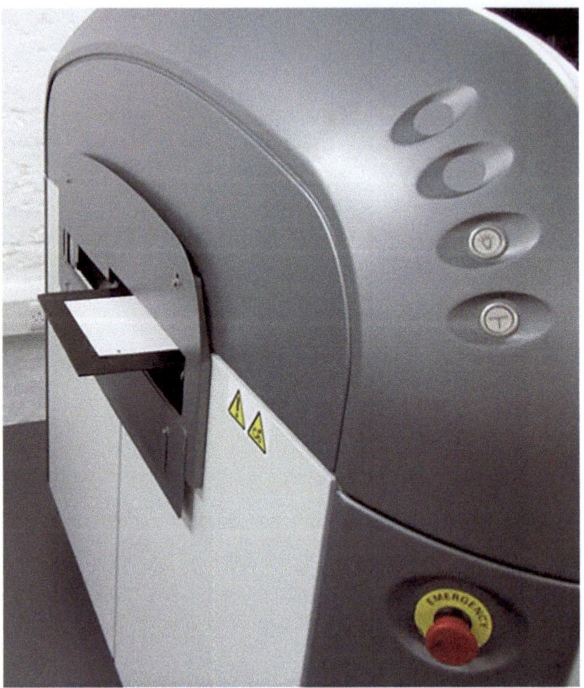

10. Now the screen-print cycle can occur, select run and click the 'GO' button, upon this instruction the machine will pull the substrate into the machine and proceed with the cycle. Firstly, the flood bar lowers and then drags the ink over the mesh design, the squeegee will then return to the starting position applying a pressure which forces the ink to pass through the stencil design.

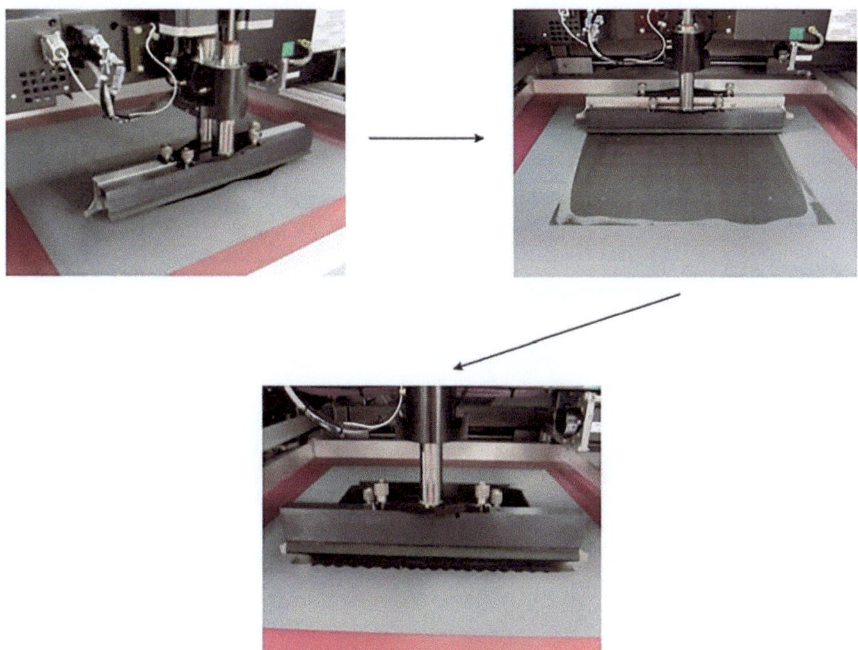

11. When the cycle has finished the substrate will be released by the machine, with
    the stencil design transferred on the top of it. A thorough inspection should
    indicate any unprinted areas or regions of uneven coverages. If the print is com-
    promised consult the troubleshooting section.

12. Repeat until the required number of electrodes has been accomplished. To complete the process, curing of the ink must occur, within an oven at a step temperature and time; usually given by the ink manufacture.

13. To increase productivity, it is good practice to clean the equipment used throughout the screen-printing process, while the electrodes cure. First, remove the squeegee and the flood bar, being careful not to damage the screen. After, remove the screen (handling it in the same fashion as previously mentioned), and place within a well-ventilated area and clean using an appropriate solvent. When cleaning the screen, you may apply slight force on the underside as this does not affect the stencil design, however you must take serious care when cleaning the top of the screen, which contains the stencil design.

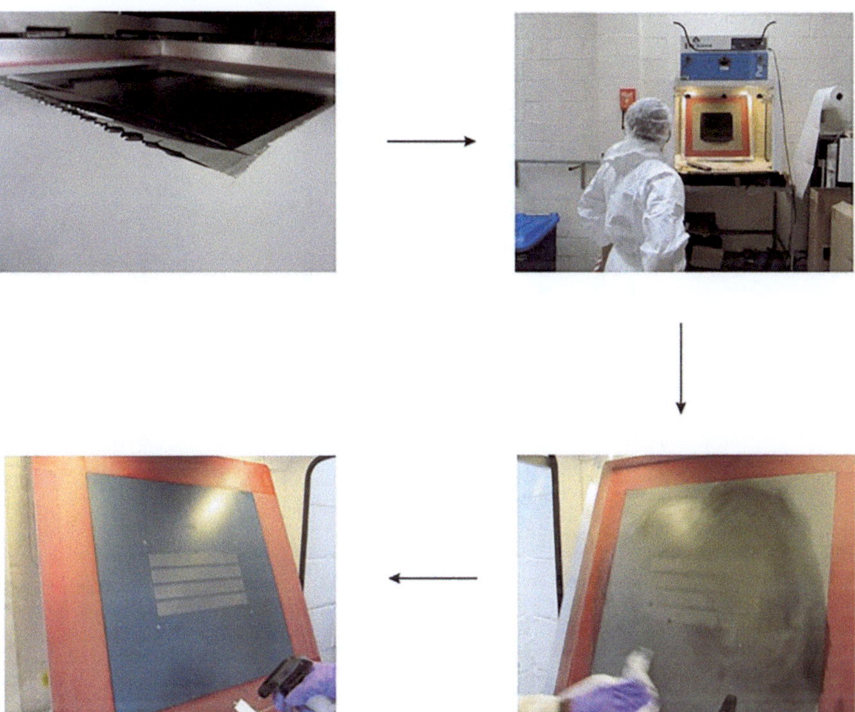

Upon completion of the first layer, movement onto the second, third and in some cases fourth can begin, utilising the required screens the layers can be produced in a reproducible and quick fashion.

3.2   Screen-Printing Troubleshooting Guide

Presented below is a detailed troubleshooting table indicating the problem, the cause and how to resolve many commons issues when utilising the screen-printing process (Tables 3.1).

Table 3.1 Troubleshooting guide for the screen-printing process

Issue	Equipment causing the problem	Cause of the problem	How to resolve issue
Screen-print incomplete	Squeegee	Not parallel with the screen	Turn blade round or renew blade
		Gap is too large between the squeegee and the screen	Decrease the gap between the squeegee and screen
		The squeegee is not wide enough for full coverage	Minimum squeegee must have a width that is 10 mm larger on each side of screen image
	Printing medium	Viscosity of printing medium too thick for successfully filling of screen design	Use a recommended screen; with a less viscous ink or emulsion
		Dried ink within the screen design	Remove the screen and clean with appropriate solvent
	Screen	Incorrect positioning of the print area	Correct so that the print design is in the middle of substrate
		Print gap too large or small	Change to the appropriate screen-substrate gap
		Screen mesh too fine	Change to screen recommended for work
Erroneous and undesirable screen-print	Squeegee	Areas of unprinted ink, could be caused by dusty/ dirty components	Raise print head and clean the underside removing any dirty or dust
		Image may contain streaks check squeegee edge for worn/damaged behaviour	Fit new squeegee blade
		Printing medium not cleared from the screen mesh due to the squeegee pressure being too low	Increase pressure in small steps until a clear track is obtained and then increase slightly

Table 3.1 (continued)

Issue	Equipment causing the problem	Cause of the problem	How to resolve issue
		Print smudging from the squeegee to screen gap being too small	Increase the gap until the screen peels away from substrate in a controlled manner
		Squeegee pressure too heavy causing poor definition of the edge of the design	Reduce squeegee pressure in small steps until good print is obtained, then increase pressure slightly
	Printing medium	Stringing and serrating of the image edges due to the ink being too thin	Use thicker or more viscous inks
		Thin deposit of ink upon the print	Use less viscous ink
	Screen	Mesh damaged and affecting the print area	Fit a new screen
		Print image serrated at edge, and has poor definition as the screen mesh is too coarse or the screen open area is too small	Change to screen to one with a recommended finer mesh/use recommended screen with thicker emulsion
		Deposit too thin, from the screen mesh being too fine	Use a coarser mesh
Substrate issues	Substrate	Contaminated printing surfaces or excessively bowed	Clean substrates thoroughly before printing. Increase vacuum, mechanical gripping, reduce squeegee pressure or reject substrate
		Vacuum is insufficient to hold substrates when very viscous pastes are used	Check whether: 1. Vacuum pipe not fully secured to unions 2. Holes at registered printing position are blocked 3. Filter of pump is blocked 4. Pump exhaust pipe is restricted 5. Other leaks exist
		Print gap too small	Set print gap correctly

Chapter 4
Quality Control/Quality Assurance Analysis of Electrochemical Screen-Printed Sensors

Contents

4.1 Cyclic Voltammetry

This technique is the most widely utilised for obtaining quantitative information about electrochemical reactions. Cyclic voltammetry consists of scanning (linearly) the potential of the working electrode using a triangular potential wave form (Fig. 4.1). The potential is swept from E_1 to E_2 and the rate at which this is achieved is the voltammetric scan rate (or the gradient of the line), as shown in Fig. 4.1 (V/s). In this case, if the potential is stopped at E_2, this is known as a linear sweep experiment. If the scan is returned back to E_1, a full potential cycle, this is known as cyclic voltammetry. Depending on the information sought, either single or multiple cycles can be performed. For the duration of the potential sweep, the potentiostat measures the resulting current that arises via the applied voltage (potential). The plot of current versus potential (voltage) is termed a 'cyclic voltammogram'. A cyclic voltammogram provides insights into redox potential, reaction kinetics, diffusion coefficient and mechanisms of electrode transfer.

Figure 4.2 shows a typical cyclic voltammetric curve for the case of the electrochemical process: $A(aq) + e^- \rightleftarrows B(aq)$, where A is the species that is dissolved into

© The Author(s), under exclusive license to Springer Nature Switzerland AG 2026

C. E. Banks, R. D. Crapnell, *Screen-Printing Electrochemical Architectures*, SpringerBriefs in Applied Sciences and Technology, https://doi.org/10.1007/978-3-032-10411-3_4

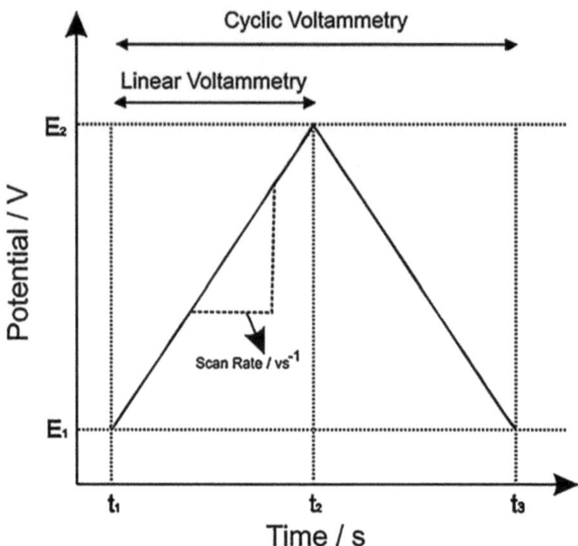

Fig. 4.1 Potential–time profiles used to perform linear sweep and cyclic voltammetry

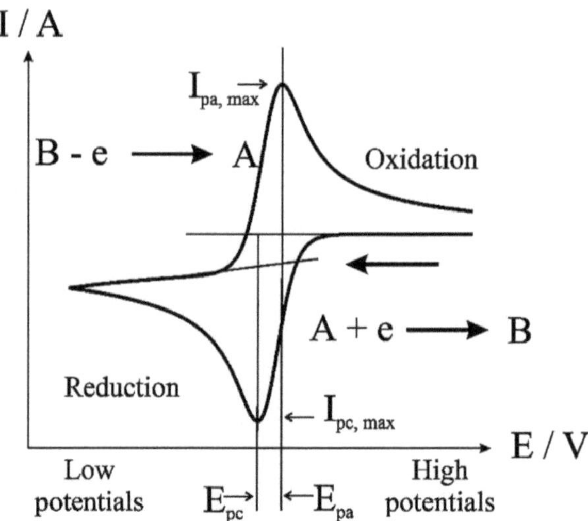

Fig. 4.2 Typical reversible cyclic voltammogram using an ideal outers-here redox probe depicting the peak position E_P and peak height I_P

an electrolyte (e.g. 1 M KCI) and will be electrochemical reduced, by accepting an electron, e^-, where B is the product. This is for an outer-sphere redox probe (see Sect. 4.4.). Note that characteristics of the voltammogram which are routinely monitored and reported are the peak height (I_P) and the potential at which the peak occurs (E_P). The magnitude of the cyclic voltammogram and E_P can be utilised for quality control purposes, to verify that a screen-printed batch conforms to an expected output; those outside this range can be discarded.

The cyclic voltammetry rises from the flux of electroactive species to the electrode surface which is governed by the Nernst–Planck equation:

$$j_j(x) = -D_j \frac{C_j(x)}{\partial x} - \frac{z_j F}{RT} D_j C_j \frac{\partial \phi(x)}{\partial x} + C_j v(x) \tag{4.1}$$

where this is comprised of three terms:

1. Diffusion, $-D_j \dfrac{C_j(x)}{\partial x}$, follows Fick's law, where D_j is the diffusion coefficient $(\mathrm{cm^2\ s^{-1}})$, $C_j(x)$ is the concentration of species j at position x.

2. Migration (due to electric field), $\dfrac{z_j F}{RT} D_j C_j \dfrac{\partial \varnothing(x)}{\partial x}$, which described the movement of charged species in response to an electric potential gradient where Z_j is the charge number of species j, and $\partial \varnothing(x)/\partial x$ is the potential gradient along the x-axis.

3. Convection, $C_j v(x)$ which described the bulk movement of species j and $v(x)$ is the rate with which a volume element moves in solution know as convection or bulk flow.

Under most cyclic voltammetry experiments, diffusion dominates since these are performed under stagnant conditions, where convection, $C_j v(x)$ is negligible and the migration (electric field), $-\dfrac{z_j F}{RT} D_j C_j \dfrac{\partial \varnothing(x)}{\partial x}$ is suppressed by adding in a supporting electrolyte (e.g., 1 M KCl) which minimises the electrode field effects.

Note that under cyclic voltammetry conditions, eq. 4.1 reduces to Fick's first law in a one-dimensional system:

$$j_j(x) = -D_j \frac{\partial C_j(x)}{\partial x} \tag{4.2}$$

which shows that flux is proportional to the negative of the concentration gradient. Species move from high to low concentration. This is a steady-state expression: it gives how much material flows per unit area but does not tell you how concentrations evolve in time. The second Ficks law is:

$$\frac{\partial C_j(x,t)}{\partial x} = D_j \frac{\partial^2 C_j(x,t)}{\partial x^2} \tag{4.3}$$

which relates how the concentration changes with time due to diffusion. The above equations are used together in modelling electrochemical cells where the Fick's law described how species move due to concentration gradients. The Nernst equation is given by:

$$E = E_f^0 \left(\frac{A}{B}\right) + \left(\frac{RT}{nF}\right) In\left(\frac{[B]}{[A]}\right) = E_f^0(A/B) + 2.3026\left(\frac{RT}{nF}\right)\log_{10}\left(\frac{[B]}{[A]}\right) \tag{4.4}$$

where E_f^0 is the formal potential, F is Faraday's constant, R is the universal gas constant, n is the number of electrons, and T is the temperature. The Nernst equation is used to describe the equilibrium potential focusing on thermodynamics, using concentrations to determine the potential. Returning to the electrochemical process described above, electrochemical rate constants for the one electron oxidation of:

$$A(aq) + e^- (m) \underset{k_c}{\overset{k_a}{\rightleftarrows}} B(aq) \tag{4.5}$$

where the k_c and k_a describe the rate constant for cathodic and anodic processes. Furthermore, we know that the net rate (flux) of reaction is given by:

$$j = k_c [A]_0 - k_a [B]_0 \tag{4.6}$$

using the above we can write:

$$j = k_c^0 \exp\left[\frac{-\alpha F\left(E - E_f^0\right)}{RT}\right][A]_0 - k_a^0 \exp\left[\frac{(1-\alpha)F\left(E - E_f^0\right)}{RT}\right][B]_0 \tag{4.7}$$

noting that the flux, $j = i/FA$, which makes:

$$i = FA\left(k_c^0 \exp\left[\frac{-\alpha F\left(E - E_f^0\right)}{RT}\right][A]_0 - k_a^0 \exp\left[\frac{(1-\alpha)F\left(E - E_f^0\right)}{RT}\right][B]_0\right) \tag{4.8}$$

In the case of a dynamic equilibrium, where the oxidation and reduction currents exactly balance each other and since no net current flows, J = 0 and the fact that $\alpha = 0.5$, gives:

$$E = E_f^0 + \frac{RT}{F} \mathit{In}\left(\frac{[B]}{[A]}\right) + \frac{RT}{F} \mathit{In}\left(k_a^0 / k_c^0\right) \tag{4.9}$$

where no net current flows the potential is given by:

$$E = E_f^0 + \frac{RT}{F} \mathit{In}\left(\frac{[B]}{[A]}\right) \tag{4.10}$$

so that $k_{ox}^0 = k_{red}^0 = k^0$ which becomes the Nernst equation. Therefore, we can write:

$$k_{red} = k^0 \exp\left[\frac{-\alpha F\left(E - E_f^0\right)}{RT}\right] \tag{4.11}$$

$$k_{ox} = k^0 \exp\left[\frac{(1-\alpha)F\left(E - E_f^0\right)}{RT}\right] \tag{4.12}$$

which are the most accessible forms of the *Butler–Volmer* expression for the electrochemical rate constants k^0_{red} and k^0_{ox}. Simply put, the Nernst equation tells you the equilibrium potential based on concentrations while the Butler–Volmer equation tells you how fast the reaction proceeds when you're not at equilibrium.

In a reversible system, as shown in Fig. 4.2, the electron transfer is fast enough that the concentrations of oxidised and reduced species at the electrode surface instantly adjust to changes in applied potential. If we consider the case of the reversible cyclic voltammogram (Fig. 4.2) at positive potentials no current flows, as the potential is driven negatively, the electrochemical rate constant (k^0) becomes large enough and the current rises as the potential becomes increasingly negative. Last, at the more negative potential the current passes a maxima and decreases. Readers are directed to the following textbooks for a rigorous analysis of cyclic voltammetry [1]. In the 'reversible' limit the electrode kinetics are so 'fast' (relative to the rate of mass transport) that Nernstian equilibrium is attained at the electrode surface throughout the voltammogram with concentrations of A and B at the electrode surface governed by the Nernst equation. During a potential sweep, electroactive species are consumed (or generated) at the electrode surface, establishing a concentration gradient that evolves into a time-dependent diffusion layer which is dependent on the rate of conversion of A to B. This depends on the formal potential $(E - E^0_f)$ of the redox probe and the heterogeneous electron transfer (k^0), where the rate of diffusion is governed by Fick's first law (Eq. 4.2), and the diffusion layer, δ, is given by:

$$\delta = \sqrt{Dt} \tag{4.13}$$

where t is given by the potential scan rate (ν): $t = RT/F\nu$ which leads to:

$$\delta = \sqrt{RTD / F\nu} \tag{4.14}$$

The concentrations of oxidised and reduced species at the electrode interface are not static but are dictated by this evolving diffusion profile. These local concentrations determine the equilibrium potential through the Nernst equation, establishing the thermodynamic reference for the system, while simultaneously entering the Butler–Volmer formalism as key parameters governing the forward and reverse charge-transfer rates. Under conditions of rapid mass transport, the interfacial concentrations closely approximate bulk values, yielding Nernstian, reversible behaviour characterised by minimal peak-to-peak separations. By contrast, when diffusion is rate-limiting, significant depletion or accumulation occurs at the electrode interface, leading to non-ideal voltammetric responses, increased peak separations, and current magnitudes governed by diffusion-controlled fluxes; this explains why the Butler–Volmer predicts that the flux of species to rise exponentially when in reality, it is mass transfer limited.

Figure 4.3 shows the response of a quasi-reversible and an irreversible cyclic voltammograms when one can observed the peak-to-peak separation increases.

Figure 4.3 shows the cyclic voltammetric response for an irreversible electrochemical couple (dotted line) where appreciable over-potentials are required to

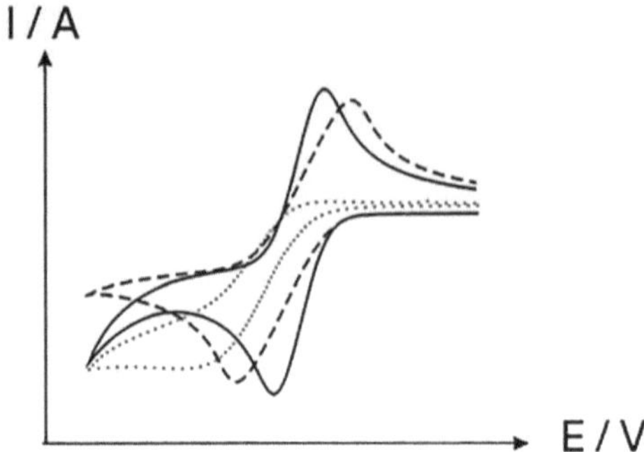

Fig. 4.3 Cyclic voltammograms for reversible (*solid line*), quasi-reversible (*dashed line*) and irreversible (*dotted line*) electron transfer

drive the reaction, as evidenced by the peak height occurring at a greater potential than that seen for the reversible case. The magnitude of the voltammetric current observed at a planar macroelectrode is governed by the following a reversible Randles–Ševćik equation:

$$I_p^{\mathrm{Rev}} = \pm 0.446 n^{3/2} FA_{real} C_{bulk} \left(FD\upsilon / RT \right)^{1/2} \tag{4.15}$$

where F is the Faraday constant, υ is the applied voltammetric scan rate, C_{bulk} is the redox probe concentration, R is the universal gas constant, T is the temperature in Kelvin, A_r is the electroactive area of the electrode, and D is the diffusion coefficient, where the $\pm$ sign is used to indicate an oxidation or reductive process respectively, although the equation is usually devoid of such sign. This equation can be represented as:

$$I_p^{\mathrm{Rev}} = \pm 2.69 \times 10^5 n^{3/2} D^{1/2} A_{real} C_{bulk} \upsilon^{1/2} \tag{4.16}$$

when the solution is at 298 K. To ensure that the cyclic voltammetric response corresponds to reversible voltammetry, please consider the waveshape of the forward peak. For a reversible system:

$$E_p - E_{1/2} = 2.218 RT / F \tag{4.17}$$

$$E_{1/2} = \left(E_{p,anodic} + E_{p,cathodic} \right) / 2 \tag{4.18}$$

where the peak current occurring at potential, E_p and $E_{1/2}$ is half the peak current. Also, in the case of a reversible system, the peak-to-peak separation, ΔE_p is *independent* of the scan rate applied and the ratio between oxidation and reduction peak currents is unity: $I_p^{ox} / I_p^{red} = 1$. Furthermore, quasi-reversible or an irreversible system, for a simple one electron transfer process, given by:

$$I_p^{quasi} = \pm 0.436 n^{3/2} FA_{real} C_{bulk} \left(FD\upsilon / RT \right)^{1/2} \tag{4.19}$$

and

$$I_p^{Irrev} = \pm 0.496 n \left(n' + \alpha\right)^{\frac{1}{2}} FA_{real} C_{bulk} D^{1/2} \left(FD\upsilon / RT\right)^{1/2} \tag{4.20}$$

where α is the transfer coefficient, n is the total number of electrons transferred and n' is the number of electrons transferred before the rate determining step. In such cases, the peak-to-peak separation, ΔE_p is *dependent* on the scan rate applied and the waveshape can be described by:

$$E_p - E_{1/2} = 1.857 RT / \alpha F \tag{4.21}$$

For a reductive process, at 298 K becomes:

$$E_p - E_{1/2} = 47.7 / \alpha \ \ mV \tag{4.22}$$

where for an oxidative process:

$$E_p - E_{1/2} = 1.857 RT / \beta F \tag{4.23}$$

In the case of reversible and irreversible behaviour, where for the former indicates that these are fast electrode kinetics but for the latter, these are termed slow electrode kinetics, but in relation to what? The answer is the mass transport coefficient, m_T, given by:

$$m_T = \sqrt{D / \left(RT / F v\right)} \tag{4.24}$$

For a reversible case, $k^0 \gg m_T$, while for the irreversible case, $k^0 \ll m_T$. The use of the parameter, Λ, is given by: [2].

$$\Lambda = k^0 / \left(FD\upsilon / RT\right)^{1/2} \tag{4.25}$$

where the following ranges for Λ as follows:
Reversible: $\Lambda \geq 15$ $k^0 \geq 0.3 v^{1/2.}$
Quasi-reversible: $15 > \Lambda > 10^{-3}$ $0.3 v^{1/2} > k^0 > 2 x 10^{-5} v^{1/2}$
Irreversible: $\Lambda \leq 10^{-3}$ $k^0 \leq 2 x 10^{-5} v^{1/2}$
where $T = 298$ and $\alpha = 0.5$.
The observed electrochemical behaviour is influenced by the applied voltammetric scan rate, whereas the scan rate varies, the diffusion layer thickness changes significantly, with slow scan rates resulting in a thicker diffusion layer, while faster scan rates produce a thinner one.

Since the nature of the electrochemical process, whether reversible or irreversible, depends on the interplay between electrode kinetics and mass transport, higher scan rates tend to promote electrochemical irreversibility.

4.2 Heterogeneous Electron Transfer

At macroelectrodes the Nicholson method is routinely used to estimate the observed standard heterogeneous electron transfer rate for quasi-reversible systems using the following equation: [3]

$$\psi = k^0 \left[\pi D_0 n\upsilon F \right]^{-1/2} \left(D_r / D_0 \right)^{-\alpha/2} \tag{4.26}$$

where ψ is the kinetic parameter and is tabulated at a set temperature for a one-step, one electron process as a function of the peak-to-peak separation (ΔE_P), where one determines the variation of ΔE_P with t and from this, the variation in the ψ. To avoid the use of the table, see the relation between ΔE_p against ψ, providing a simple equation from which ψ can be readily deduced: [4]

$$\psi = \left(-0.6288 + 0.0021X \right) / \left(1 - 0.017X \right) \tag{4.27}$$

where X indicates that ΔE_p x n, leading to:

$$\psi = \left(-0.6288 + 0.0021\Delta E_p \right) / \left(1 - 0.017\Delta E_p \right) \tag{4.28}$$

when the peak separation, ΔE_p is greater than 212 mV, one can use the following approach: [5]

$$k^0 = \left(2.18\alpha D\pi nF / RT \right)^{1/2} \exp\left[\alpha^2 nF / RT \right]\Delta E_p \tag{4.29}$$

It is useful to know the generic Randles–Ševćik equation for reversible behaviour:

$$I_p = -\zeta\left(p \right)\sqrt{\frac{n^3 F^3 \upsilon D}{RT}}A_{real}C_{bulk} \tag{4.30}$$

where $p = r\sqrt{\dfrac{nF\upsilon}{RTD}}$

for the case of different electrochemical geometries:

1. Planar disc electrode: r = radius, $\zeta(p)$= 0.446 which is the Eq. 4.15;
2. Spherical or hemispherical electrode: r = radius, $\zeta(p)$= 0.446 + 0.752p^{-1}.
3. For a small disk electrode: r = radius, $\zeta(p)$= 0.446 + 0.840 + 0.433$e^{-0.66p}$ —0.16 6$e^{-11/p}$)p^{-1} ~ 0.446 + 4/pp^{-1}
4. For a cylinder or hemi-cylinder: r = radius, $\zeta(p)$= 0.446 + 0.344$p^{-0.852}$.
5. For a band electrode: $2r$ = width, $\zeta(p)$= 0.446 + 0.614(1 + 43.6p^2)$^{-1}$ + 1.323$p^{0.892}$ ~ 0.446 + 3.131$p^{-0.892}$

The above analysis allows one to readily determine key parameters (such as E_p, I_p, k^0, etc.) with the utilisation of screen-printed electrodes.

4.3   Selection of Redox Probes

In the above case, we considered the generic redox example. To run the cyclic voltammetric experiment described above, one needs to select an appropriate electroactive analyte. Typically, these probes are classified into inner- and outer-sphere redox probes. A wide range of redox probes have been investigated on carbon-based electrodes such as highly ordered pyrolytic graphite (HOPG) and glassy carbon (GC), including: $IrCl_6$, $Ru(NH_3)_6$, $Co(phen)_3$, MV, $Fe(phen)_3$, $Fe(CN)_6$, $Co(en)_3$, $Ru(en)_3$, $Fc(COOH)_2$, $Ru(bpy)_3$, $Ru(NH_3)_5py$, $Co(sep)$, $Ru(CN)_6$, $Mo(CN)_8$, $W(CN)_8$, $Fe_{(aq)}$, $Eu_{(aq)}$, and $V_{(aq)}$. Outer-sphere redox probes are considered surface-insensitive, in that the standard heterogeneous rate constant is largely unaffected by the oxygen-to-carbon ratio of the surface, the degree of cleanliness, or the presence of neutral adsorbates forming monolayer films. In these systems, no specific adsorption or chemical interaction with surface groups occurs; instead, the electrode simply functions as an electron source or sink. Consequently, the electrochemical response of outer-sphere probes depends primarily on the electrode's electronic structure, particularly the density of electronic states near the Fermi level. By contrast, inner-sphere redox probes are inherently surface-sensitive. Their electron transfer kinetics are strongly influenced by the electrode surface chemistry and microstructure, with catalytic interactions that can be significantly inhibited by adsorbates, impurities, or passivation layers. Such systems often require specific interactions with oxygenated surface species, which may enhance or hinder the electron transfer process depending on the chemical environment. In these cases, the electrochemical response is dominated by surface functionality and adsorption phenomena, rather than by the electronic density of states. Moreover, inner-sphere redox couples typically exhibit higher reorganisation energies, further emphasising their dependence on surface-mediated interactions.

The contrasting behaviour of inner- and outer-sphere redox systems provide a powerful diagnostic tool for probing electrode surface chemistry and structure. As emphasised by McCreery, a systematic framework or 'road map' of commonly employed redox probes (Fig. 4.4) enables researchers to identify whether observed electrochemical responses arise primarily from surface-state effects or from intrinsic electronic properties of the electrode material.

When using surface-sensitive probes and/or one has introduced significant C-O moieties into the electrochemical surface, for example, see Sect. 1.6, one might observe adsorbed voltammetry, where symmetrical oxidation and reduction cyclic voltammetric profiles will be observed. The peak current is related to the surface coverage (Γ) and potential scan rate for a reversible process:

$$I_p = n^2 F^2 \Gamma A \upsilon / 4RT \tag{4.31}$$

where the charge (Q) is related to the surface coverage by the following expression:

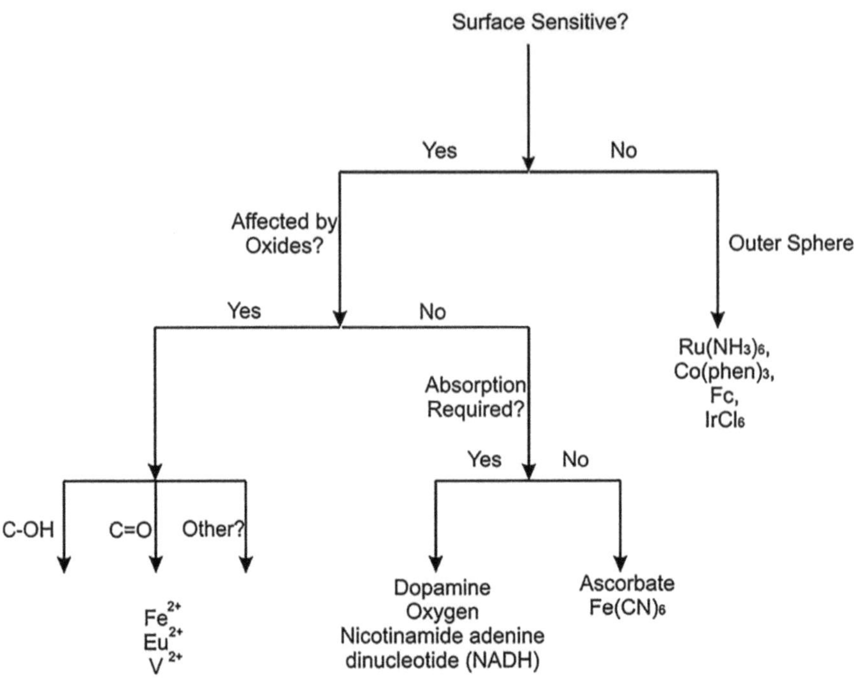

Fig. 4.4 Classification of redox systems according to their kinetic sensitivity to particular surface modifications on carbon electrodes. Adapted from Ref. 6

$$Q = nFA\Gamma \tag{4.32}$$

The diagnosis of an adsorbed species is to explore the effect of scan rate on the voltammetric response, which should yield a linear response for the case of peak current against scan rate where a plot of $\log_{10}$ (peak current) against $\log_{10}$ (scan rate) was found to linear with a slope of 1.0 (or close to). That said, you may have a weakly adsorbed species where the slope may be from 0.7 to 1.0. In the case that you have a strongly adsorbed species undergoing irreversible redox reaction: [7]

$$E_p = E^{0'} - \left(\frac{RT}{\alpha nF}\right)\ln\left(\frac{RTk_s}{\alpha nF}\right) + \left(\frac{RT}{\alpha nF}\right)In\upsilon \tag{4.33}$$

where the above equation can be used to determine the number of electrons in the electrochemical process assuming that $\alpha = 0.5$. Furthermore, the heterogeneous electron transfer rate constant (k_s) constant can be deduced from:

$$\log k_s = \alpha Log(1-\alpha) + (1-\alpha)\log\alpha - \log(RT/nF\upsilon)$$
$$-\left(\alpha(1-\alpha)nF\Delta E_p / 2.3RT\right) \tag{4.34}$$

which is valid for n x $\Delta E_p > 200$ mV.

4.4   Changing the Electrode Geometry: Macro to Micro

As mentioned in Chap. 1, macro- and micro-electrodes can be fabricated via screen-printing. At a macroelectrode, electrolysis of A occurs across the entire electrode surface such that the diffusion of A to the electrode or B from the electrode surface is termed planar, and the current response is typically described as 'diffusion limited', giving rise to an asymmetric peak as shown in Fig. 4.5a. At the edge of the macroelectrode, where the electrode substrate meets the insulting material defining the electrode area, diffusion to or from the edge of the electrode is effectively to a point. Therefore, the flux, j, and the rate of mass transport are larger at the edge and as such diffusion becomes convergent. This is termed an 'edge effect' which is negligible at a macroelectrode since the contribution of convergent diffusion to the edges of the macroelectrode is inundated by that of planar diffusion to the entire electrode area.

As the electrode size is reduced from macro to micro, or even smaller to that of nano, convergent diffusion to the edges of the electrode becomes significant. In this regime a change in the observed voltammetric profile is observed which results in the loss of the peak shaped response, as evident in Fig. 4.5b with that of a sigmoidal voltammogram. The effect of convergent diffusion has the benefit of improvements in mass transport such that the current density is greater than at a macroelectrode under planar diffusion (Fig. 4.5).

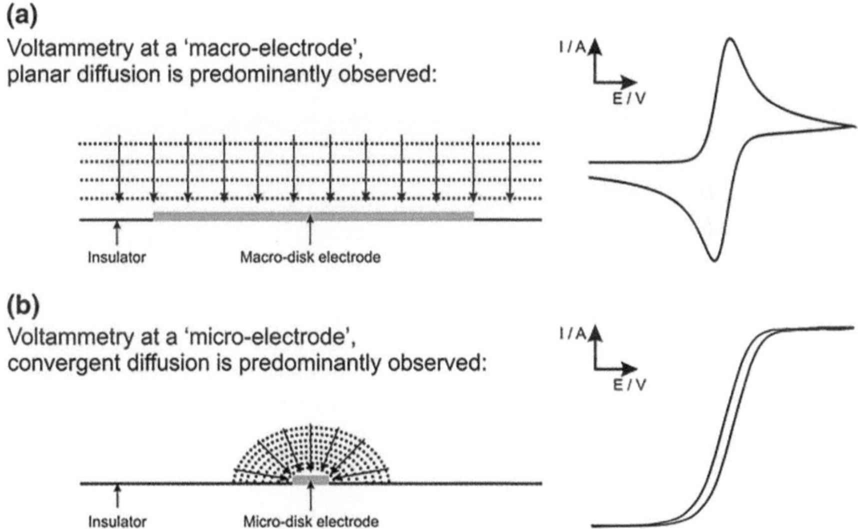

Fig. 4.5 The unique differences between the cyclic voltammetric signatures observed at a macroelectrode (**a**) compared to a microelectrode (**b**)

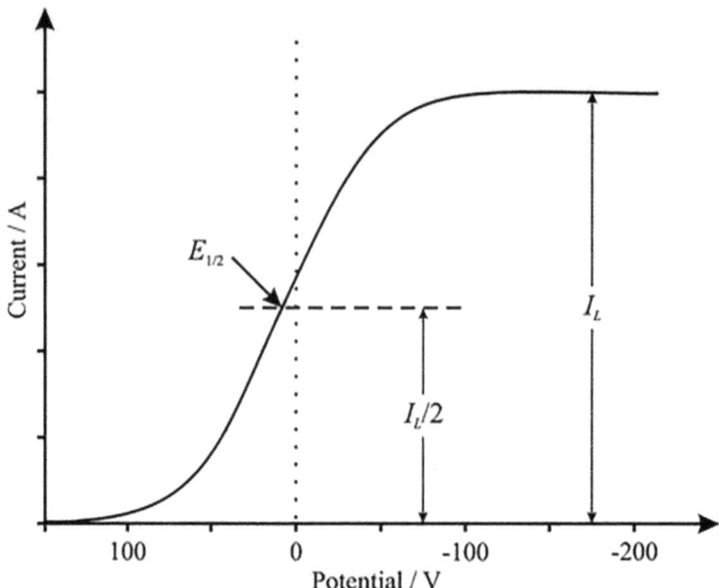

Fig. 4.6 Steady-state voltammogram for a reversible process observed at a microelectrode

For a reversible electrode reaction at a microelectrode, as shown in Fig. 4.6, where $E_{1/2}$ is the half-wave potential, the following equation describes the expected voltammetric shape:

$$E = E_{1/2}^{rev} + RT / nFIn\left(I_L - I / I_L\right) \tag{4.35}$$

where $E_{1/2}^{rev} = E^{0'} = RTnFInD_R^{1/2} / D_0^{1/2}$ since the ratio of diffusion coefficient is nearly equal, $E_{1/2}$ is a good approximation for $E^{0'}$ for a reversible couple. When a plot of E against $In(I_L - I/I_L)$ is constructed, a linear response should be1 2 observed with a gradient equal to RT/nF and an intercept of $E_{1/2}^{rev}$ if the waveshape corresponds to a reversible process. The effect of different electrochemical kinetics is known for the limiting current to decrease and the $E_{1/2}$ shifts. To determine between reversible, quasi-reversible and irreversible, a useful approach is the Tomeš criteria [8].

4.5   Surface Characterisation

It is vital when utilising such screen-printed architectures or any electrochemical setup that the visual and surface characterisation of such system is analysed, therefore informing the user on the true surface of the electrode. A common technique for this is scanning electron microscopy (SEM) which allows the user to analyse the surface and produce images of the areas that could be potentially interesting. Shown within Fig. 4.7 is a typical optical and SEM image of the screen-printed electrodes,

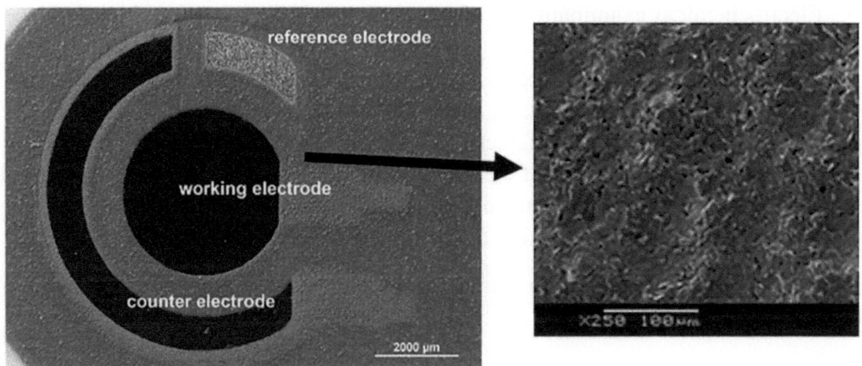

Fig. 4.7 Optical image of a graphitic screen-printed electrode. Figure from [9]. Open Access. SEM image of a graphitic screen-printed electrode. (Figure from [10])

it is visible that there are large areas comprising of a range of carbon/graphitic material, solvents and polymeric binders. The latter are electro- chemically inactive and will change the true electrochemically active area. On the other hand, due to the way the inks are formulated and depending on the quality of the print, the electrode might be significantly rough.

Another useful approach is to sample a random number from a batch to try and determine whether the overall batch conforms to set standards. In addition to the aforementioned methods, an optical approach can be carried out to map the electrode surface; this can be used to define the diameter and circumference of the electrode allowing the fabricator to reject sensors that do not meet the criteria expected. However, this will not give an indication of the electrode roughness.

A method to measure the surface roughness which can potentially be used in QC/QA analysis of the fabricated electrodes is white light profilometry.

From this technique, a surface roughness profile can be achieved measuring the surface topography of the screen-printed electrode to determine a roughness factor (R_F) and are represented in a surface map of the electrode. The surface profile maps can be analysed using a Matlab script or similar software to work out the surface topology measurements. This approach also allows for the analysis of the thickness of a deposit from the resulting screen-print, as utilising different inks will give varying thicknesses. It is noted that if the roughness factor starts to change significantly through a batch it could be the case that the ink requires a greater agitation prior to printing.

4.6   Quality Control of Screen-Printed Electrodes

In summary the quality control of these systems can include:

- A visual inspection for print quality.
- In-process inspection checks such as the resistance measurement of the dried printed structure, image stretch and layer-to-layer registration measurements.

- Electrochemical examination (cyclic voltammetry) across printed card, row to row and through batch measuring *Ip*, *Ep* and k^0. Also, the use of resistance using a multimeter and EIS, see Sect. 1.5 [11]. EIS is not considered further but readers are directed to: [12, 13] Last, the use of XPS is recommended to deduce the C-O moieties and C/O ratios (see Sect. 4.3).
- Microscopic measurement of the diameter and/or geometric area of the dried ink deposit upon the screen-printed electrode using a coordinate measuring machine.
- The measurement of deposition thickness via white light profilometry.

A combination of the above can be used to determine the quality of a batch of screen-printed electrodes.

References

1. R.G. Compton, C.E. Banks, *Understanding Voltammetry*, 4th edn. (World Scientific Publisher, 2025)
2. H. Matsuda, Y. Ayabe, Zeitschrift für Elektrochemie, Berichte der Bunsengesellschaft für physikalische Chemie **59**, 494–503 (1955)
3. R.S. Nicholson, Anal. Chem. **37**, 1351–1355 (1965)
4. I. Lavagnini, R. Antiochia, F. Magno, Electroanalysis **16**, 505–506 (2004)
5. R.J. Klingler, J.K. Kochi, J. Phys. Chem. **85**, 1731–1741 (1981)
6. K.K. Cline, M.T. McDermott, R.L. McCreery, J. Phys. Chem. **98**, 5314–5319 (1994)
7. E. Laviron, J. Electroanal. Chem. Interfacial Electrochem. **101**, 19–28 (1979)
8. T. J, Coll Czech Chem Comm **9**, 12–21 (1937)
9. K. Tyszczuk-Rotko, J. Kozak, B. Czech, Sensors **22**, 2437 (2022)
10. P.M. Hallam, D.K. Kampouris, R.O. Kadara, C.E. Banks, Analyst **135**, 1947–1952 (2010)
11. M.J. Whittingham, N.J. Hurst, R.D. Crapnell, A. Garcia-Miranda Ferrari, E. Blanco, T.J. Davies, C.E. Banks, Anal. Chem. **93**, 16481–16488 (2021)
12. E.P. Randviir, C.E. Banks, Anal. Methods **14**, 4602–4624 (2022)
13. L.M. Peter, *Electrochemical Impedance Spectroscopy and Related Techniques from Basics to Advanced Applications* (World Scientific Publisher, 2024)